T0127630

THE FRONTIERS COLLECTION

THE FRONTIERS COLLECTION

The books in this collection are devoted to challenging and open problems at the forefront of modern science and scholarship, including related philosophical debates. In contrast to typical research monographs, however, they strive to present their topics in a manner accessible also to scientifically literate non-specialists wishing to gain insight into the deeper implications and fascinating questions involved. Taken as a whole, the series reflects the need for a fundamental and interdisciplinary approach to modern science and research. Furthermore, it is intended to encourage active academics in all fields to ponder over important and perhaps controversial issues beyond their own speciality. Extending from quantum physics and relativity to entropy, consciousness, language and complex systems—the Frontiers Collection will inspire readers to push back the frontiers of their own knowledge.

More information about this series at
http://www.springer.com/series/5342

Yair Neuman

How Small Social Systems Work

From Soccer Teams to Jazz Trios and Families

 Springer

Yair Neuman
Department of Cognitive and Brain Sciences
Ben-Gurion University of the Negev
Beersheba, Israel

ISSN 1612-3018   ISSN 2197-6619   (electronic)
THE FRONTIERS COLLECTION
ISBN 978-3-030-82240-8   ISBN 978-3-030-82238-5   (eBook)
https://doi.org/10.1007/978-3-030-82238-5

This Springer imprint is published by the registered company Springer Nature Switzerland AG
The registered company address is: Gewerbestrasse 11, 6330 Cham, Switzerland

This book is dedicated to my son, Yiftach, and his wife, Shir—a small, happy system that I hope will play jazz by carefully attuning to each other's perspective.

Preface

When I was a graduate student, chess was considered the ultimate expression of intelligence and the real benchmark for testing artificial intelligence. However, in what may retrospectively be considered an act of heresy, I proposed a new criterion for game intelligence and wrote an "insight" paper explaining why soccer is a much more intelligent game than chess. I shared the paper with one of my esteemed professors, who described it as "wise" but was highly critical of its thesis. At that time, soccer was considered to be no more than a "group of hooligans kicking a ball" (see Chap. 2) and the idea of studying soccer in the academy was not welcomed to say the least. The best analogy I can use to explain this past *Zeitgeist* in which the study of soccer was not welcomed may be a gentleman participating in a bareknuckle fight—an unthinkable scenario that contradicts the essence of a civilized person. Soccer was seen as the ultimate entertainment of the mob and no more.

It took me more than 20 years to return to soccer as a case study for a complex adaptive system. Through this case study, I have realized how little we really understand small systems, from soccer teams to families, and this awareness has sparked my interest in understanding small social systems. Our deficiency in understanding small systems may seem surprising, but think about sport commentators. If you listen to sport commentators, you may be entertained by their interpretations, insights, "understanding," and "knowledge"—all expressed with high confidence and amusing rhetoric. However, when you pay close attention to the commentaries, you learn that they are no more than entertaining chats that add nothing to our understanding of the match.

Surprisingly, you may also find the same shortcoming in our understanding of other small systems, such as families. This shortcoming is expressed in the effectiveness or more accurately in the lack of evidence, supporting the effectiveness of couples and family therapy (see, e.g., Rathgeber et al., 2019). From a scientific perspective, real understanding cannot be separated from the world and a real understanding of family dynamics should be implemented in an effective family therapy. Regardless of the theories, models, and insights developed and gained in marital and family therapy, the bottom line is that families are probably understood to the same extent as soccer teams. It therefore seems that beyond intuition and insights, family and marital therapists fail to scientifically understand families, just as sport commentators fail to scientifically understand soccer teams. There is nothing wrong with being a commentator or in working with intuition, but if we would really like to understand small systems, then a different approach is required.

The quest for this understanding motivated me to write this book. While the book has no pretentions to provide the silver bullet for understanding small social systems, it seeks to point to a different way of thinking about small systems. The book should therefore be read with this approach in mind and with the vision of moving from being a commentator to the possibility of grounded understanding, with all its real-world implications.

Beersheba, Israel   Yair Neuman

Acknowledgments

I would like to thank Yochai Cohen, Dennis Noble, Boaz Tamir, and Dan Vilenchik for collaborating with me on several papers dealing with small systems; Mario Alemi for a constructive reading of the final manuscript; Saadia Gozlan for being my expert guide to soccer; Irun Cohen for challenging discussions; My university rector, Heim Hames, for his continuous support; Springer's editor Angela Lahee, for enthusiastically supporting my project and showing good taste for innovation; and my copyeditor Hazel Bird, who has edited six of my books, including this one, with sense and sensibility.

Contents

About the Author

Prof. Yair Neuman (b. 1968) is the author of numerous papers and eight academic books published by leading publishers. He has been a visiting professor at various universities, such as Oxford, Toronto and MIT, and holds a full professorship at the Department of Cognitive and Brain Sciences, Ben-Gurion University of the Negev, Israel. His most recent book, dealing with mathematics and literature, appeared in a new book series published in collaboration with the Fields Institute for Research in Mathematical Sciences.

Part I

The Foundations of Small Social Systems

1

Introduction: On Unhappy Families, Unsuccessful Soccer Teams, and Grandmothers' Intuition

Science does not try to explain; it hardly even interprets; science places models above all.
John von Neumann

In this introductory chapter, I explain why small social systems, such as families and soccer teams, are much more complex than we might assume; briefly present the main thesis of the book; explain my general approach and my emphasis on both theoretical understanding and scientific modeling; justify the style of the book; and invite readers both expert and non-expert.

In his famous opening to the novel *Anna Karenina*, Tolstoy writes that "all happy families are alike" but that "each unhappy family is unhappy in its own way" (Tolstoy, 1877/2002, p. 1). When I shared this opening with a friend who is a scientist, he disagreed with Tolstoy and argued that the opposite is true. He said that most unhappy families he had encountered were much more alike than unalike, as they shared a simple common denominator, such as a missing parent. The opening statement of *Anna Karenina* is therefore debatable or, in the worst-case scenario, scientifically invalid. In this context, it is not quite clear why this opening is still cited after so many years. After all, Tolstoy's observation could have been formulated and empirically tested as a research hypothesis—which we could call the "Anna Karenina hypothesis." Either supporting or refuting this hypothesis could lead us to a better understanding of families (and happiness).

© The Author(s), under exclusive license to Springer Nature
Switzerland AG 2021
Y. Neuman, *How Small Social Systems Work*, The Frontiers Collection,
https://doi.org/10.1007/978-3-030-82238-5_1

To the best of my knowledge, the Anna Karenina hypothesis has not been scientifically confirmed or refuted. So, what makes the opening of *Anna Karenina* so memorable in contrast with the findings of numerous studies on families that regularly pop into the public awareness through the mass media just to be lost in a sea of oblivion the day after? To answer this question, we may turn to another one of Tolstoy's observations, which is not about families but about art.

Tolstoy suggests that the business of art, to include literature, is in making us understand that which is "incomprehensible and inaccessible in the form of argument" (Tolstoy, 1897/1995, p. X). This is a very interesting observation. First, it explains to us that the opening of *Anna Karenina* should not be considered a hypothesis that can be empirically tested and, second, it indicates that the opening expresses some kind of "artistic understanding" different from that gained via science and reason (i.e., as a form of argument). The fact that there has been no definitive conclusion regarding Tolstoy's observation may lead us to suspect that some systems, such as families, are much more difficult to understand than we might have naïvely assumed and that we still cannot fully understand them "in the form of argument." This is precisely where art comes into the picture, at least according to Tolstoy. Therefore, one possible reason that the opening of *Anna Karenina* is cited after all these years is that beyond intuition and artistic understanding, both of which I deeply respect, families are to a large extent still incomprehensible and inaccessible "in the form of argument."

If we accept von Neumann's thesis (cited at the start of this chapter) that science is about modeling, then we may conclude that we still cannot successfully model the behavior of families. In this context, Tolstoy's opening is important because it presents us with a quandary rather than an argument: Why do we debate such a trivial and simple issue as happy and unhappy families? Isn't the answer clear? And if not, why? Is it so difficult to model and understand small systems such as families?

The opening of *Anna Karenina* inevitably raises our awareness of the difficulty of understanding a family, which is a system that the overwhelming majority of us have experienced firsthand. Small social systems, such as families and soccer teams, might be misleading in their familiarity and apparent simplicity. However, to date, we lack a general, firm, and basic scientific understanding of them, and we are still short of methodologies for modeling their behavior. By this I don't mean that small social systems, such as families and soccer teams, are totally incomprehensible or that we have no knowledge of them. Such an argument is both an invalid generalization and a *straw man fallacy* that should be clearly avoided. We all know how to identify unhappy families or unsuccessful soccer teams, for instance. However, while a scientific understanding may *emerge* from common sense and deep intuition, it is supposed to go *beyond* the realm of common sense and intuition and to gain valid generalization beyond particular instances of (for example) small social systems.

As a graduate student, I took several advanced courses in psychological measurement and research methodology. One of my professors, a sharp-minded and highly critical person, used to say that the results of psychological research are either things his grandmother already knew (e.g. poor, broken, unhealthy families are probably unhappy) or results gained in the lab with no ecological validity or relevance to the real world where real human beings live, think, and act. I have all respect for grandmothers' knowledge and common sense. However, when pointing to the difficulty of understanding small systems, such as families, it seems justified to enrich our understanding and to go beyond what my wise late grandmother already knew. It is also justified to go beyond the limited scope of lab findings and beyond the study of small systems in their specific manifestations (e.g. families or soccer teams). If we seek to better understand small systems then we have to understand the general logic of small systems, the logic transcending their particular instances, whether in sport or in music. This suggestion invites us to inquire into the nature of small systems from the broadest and deepest scientific perspective.

The aim of the current book is therefore to address the challenge of progressing our understanding of small social systems by laying some theoretical scientific foundations and by proposing novel theoretical ideas that can explain what is unique about small social systems and why understanding them is still an open challenge. Beyond understanding, the book proposes several ways in which interesting aspects of small systems can be *modeled*, from the synergy of soccer players to the chemistry of romantic couples. The book therefore deals both with understanding and modeling but it has no pretension to provide a single "principle" to explain the behavior of small systems—just some theoretical thoughts, challenging speculations, and possible methodologies that may enrich our modeling and understanding of small systems across several domains.

The first part of the book presents the complexity, uniqueness, and scientific foundations of small systems. In this part, I first bring to our awareness how complex these small systems are. I also explain why such systems are unique and different from big systems, which have been intensively and successfully modeled, mainly by physics. The main thesis of the book is that small systems are specifically suited to *improvising* or creatively using and adjusting patterns in order to respond in real time to challenges by producing *tailor-made solutions*. I deeply discus the "how" and "why" of this thesis by incorporating references to "large" systems that have been studied in physics.

One point that I present in this part of the book is that the unique aspect of small systems that I would like to *theoretically* expose is the same as the one that *intuitively* evokes our curiosity and joy in observing their behavior, whether as observers of a soccer match or as the audience of a family drama as presented in the theater or on TV. We enjoy observing *tailor-made solutions formed in a flexible, adaptable and creative manner to address real-time and real-world challenges*. I explain why this behavior is observed in small systems that occupy a unique position between the behavior of populations, such as the population of gas molecules or those composing a species, and the free and anarchistic but extremely limited behavior of the individual "particle."[1]

In addition, I explain why the behavior of small systems is grounded in foundational physical–cognitive aspects of the world, mainly the need to "harvest" entropy in an irreversible world, which is deeply connected with cognitive processes characterizing all living creatures. In a nutshell, the idea is that all creatures must have the ability to reason backward in order to learn abstract patterns and use them to instigate adaptive *behavior*. However, acquiring this ability in a world governed by irreversibility is a challenging task. Memory is the solution to the lack of reversibility, and memory and (no less important) forgetting are best formed through interactions between a small number of particles that form *abstract patterns* that later must be used for *concrete actions*. The real secret sauce of small social systems is therefore the ability of a small number of particles/individuals to use abstract representations, albeit in a highly flexible and adaptive manner, through their mutual interactions. This idea is intensively discussed and explained in the later parts of this book.

I must emphasize again that while the first part of the book lays some theoretical foundations, it has no pretensions whatsoever to identify the ultimate and single "principle" through which the complexity of small systems may be resolved. The theoretical foundations aim to locate our understanding within some basic scientific ideas and commonsensical hypotheses, and they are accompanied by open questions, speculations, and provocative ideas that have a legitimate place in an academic book. What I propose is an approach rather than a principle, and the theoretical foundations aim to serve as a source of inspiration for developing hypotheses and modeling methodologies rather than as a picture of the world as it is. The main benefit of adopting this

---

[1] I use the term "particle" as a synonym for the individual in order to frame the discussion in the broadest context possible, beyond that of the humanities and social sciences.

perspective is that the way in which a small number of particles collaborate to address real-world challenges may provide us with some new directions for studying the behavior of small systems, from soccer teams to families. Indeed, the published papers from which this book has emerged cover a variety of topics from modeling the behavior of soccer teams (Neuman & Vilenchik, 2019) to modeling significant interactions in *Sex and the City* (Neuman et al., 2020).

While the first part of the book lays the theoretical foundations, the second part of the book presents various aspects of understanding and modeling particular behaviors of small systems, from a new way to model the importance of constraints in soccer to a new way of identifying unique interactions as they are expressed (for instance) in the film *As Good as It Gets* (1997), a romantic comedy featuring Jack Nicholson and Helen Hunt. The modeling approach doesn't pretend to teach us how small systems *really* work, as such a move would require us to fully understand the causal networks through which a system operates. I have no tools that would allow me to enter the mind of a soccer team or the minds of a couple during a romantic date. The modeling part aims to serve the limited task of representing and analyzing the behavior of small systems with the minimal hope of gaining some insights into the behavior of small systems.

My emphasis on a modeling approach deserves some explanation as models necessarily simplify and distort. Their benefit is in providing us with a simple map that we can use to address simple questions, rather than in offering us a full understanding. The map of the London Underground may nicely illustrate this point. This map is a model, a representation, but the model doesn't pretend to represent the *real* network of the Underground. If you are a passenger seeking to find his way in London, the map may be extremely helpful. However, if you are the chief engineer of the Underground's sewer system, the map may have little value for you in reaching the location of a leak.

In contrast with some oversimplistic perspectives, scientific models are maps that should be judged by their pragmatic rather than their "transcendental" value or logical coherence. They are modest and local, and as such of great potential value. Whether you are a confused touristic seeking his way in London, a sewer engineer directing her staff to the source of a leak, or a bacterium navigating toward tasty glucose, your maps and models, whether mental or concrete, are judged only by practical considerations and by whether they help you or not. While this emphasis might be considered overly fussy, it reveals some important nuances that—unless one is aware of them—might lead to hubris and serious conceptual difficulties and flaws.[2] Having said that, I now move on to the target readership of this book, as in writing the book I have formed my own model (or map) of the reader.

This book is aimed not only at the expert but also at the educated reader who has some academic background. While the book refers to various scientific fields and in the second part introduces novel methodologies for modeling small systems, these issues are all elaborated so as to make them *fully accessible*, with no assumption of expertise whatsoever. This doesn't mean that the book is easy to read, as it is not, but that the book is accessible to the educated reader or the academic who is ready to invest the required amount of effort. In a world where the fragmentation of knowledge has reached a very high level of resolution, it seems a must for a book seeking to study small social systems beyond their particular manifestations and across disciplines, to adopt an approach that is multidisciplinary, conceptual, and methodological. This means that the book draws on and integrates several disciplines, focuses on concepts rather than on formalism, and introduces novel modeling methodologies in a friendly manner that is accessible to the nonexpert.

The decision to target the book toward both the expert and the educated and curious reader has influenced its style, language, and resources. The book is written in a non-technical albeit scientifically rigorous and (it is hoped) entertaining way, using language that sometimes non-academically echoes real-life conversations and style. For example, in Chap. 3, in order to explain the idea of irreversibility, I cite my army drill sergeant saying that "there is no motherfucker who can stop time" (or "the flow of time"). The term "motherfucker" is my translation of a Hebrew word. It is an English-language vulgarism, here used as "a term of admiration, as in the term *badass motherfucker*, meaning a fearless and confident person."[3] The use of "censored" words such as "fuck" is done not for the sake of provocation, in which I have no interest in itself, but with the specific aim of keeping the book aligned with the real world, with all of its troubled nature. After all, it is the real world—with all its complexity, uncertainty, and noise—that I strive to understand in this book. In this context of sometimes using real-world language, it is an astonishing fact that in a culture where theatrical sadistic violence is justified by artistic freedom (see, for example, the 2019 film *The Joker* or the film or TV versions of *The Punisher*), too many self-righteous academics I have met are more bothered by the purity of language than by the "purity" of deeds. I have allowed the real, carnivalesque world to enter into this book, whenever required but to a limited extent, to remind myself and the reader

---

[2] See, for example, van Es (2020), where the use of the free energy principle is criticized.
[3] https://en.wikipedia.org/wiki/Motherfucker.

that it is the lived world that we're interested in with its unhappy families, failed soccer teams, and bareknuckle fights. Keeping the idea of the educated reader in mind, the book includes references to scientific academic papers and books but also includes references to various non-academic resources, such as Wikipedia and YouTube. In a book that aims to be accessible to the educated reader of our time, the snobbish exclusion of such resources has no justification whatsoever.

Whether all happy readers are alike and each unhappy reader is unhappy in his or her own way is a question I cannot answer. However, for the curious reader seeking to learn about small social systems, this book may be an interesting and enriching experience that facilitates entry to a happy family of curious people who find pleasure in a life of contemplation (*vita contemplativa*) that is deeply immersed in a life of action (*vita activa*).

Summary

- Small social systems, such as families and soccer teams, may be misleading in their familiarity and apparent simplicity.
- This book aims to progress our understanding of small social systems.
- The main thesis of the book is that small systems are specifically suited to improvising or creatively using and adjusting patterns in order to respond in real time to challenges by producing tailor-made solutions.
- The real secret sauce of small social systems is the ability of a small number of individuals to generate highly flexible behavior through their mutual interactions.
- This book draws upon and integrates several disciplines, focuses on concepts rather than on formalism, and introduces novel modeling methodologies in a friendly manner that is accessible to the nonexpert.

2

On the Foxes of Leicester, Underdogs, and Lady Fortuna

In this chapter, we are introduced to Leicester City Football Club, understand why it was considered to be an underdog at the beginning of the 2015–2016 season, learn from Jewish sages why forecasting is a craft perceived with suspicion, understand how collective dynamics is a source of surprise, and see why social psychology, whose *raison d'être* is the study of the individual in a social context, cannot really help gambling agencies to better predict gamblers' behavior.

Fortuna Doesn't Play Soccer

Leicester is a city located in the East Midlands region of England. If you are not British, then it is most likely that you will never have heard of it. Unlike London, the city is not a major hub of government or tourism. Unlike Oxford or Cambridge, the city is not associated with famous universities. It is not even associated with famous figures, unlike Nottingham and its local hero Robin Hood. In terms of *gross value added*, which is an economic measure of performance, the city is located far below London and lower than Nottingham and Coventry. In fact, the people of Leicester earn among the least in the UK,[1] a fact that can be explained by the relatively large proportion of immigrants (33%) making up the city's small population.

---

[1] https://www.bbc.com/news/business-44237180.

© The Author(s), under exclusive license to Springer Nature
Switzerland AG 2021
Y. Neuman, *How Small Social Systems Work*, The Frontiers Collection,
https://doi.org/10.1007/978-3-030-82238-5_2

The local soccer team—Leicester City Football Club—is an old club that was established in the nineteenth century. The Foxes, so called because of their logo, which portrays an image of this tricky creature, have never been considered one of the shining stars of the English Premier League. The names of soccer teams such as Arsenal, Chelsea, Liverpool, and Manchester United may be familiar even to those who are not soccer fans, but not Leicester. When the club participated in the 2015–2016 season of the Premier League, no one really considered it to be a favorite to win. In fact, prior to the season beginning, the betting odds of Leicester winning the Premier League were 5000–1.[2] It seems that almost no one really believed that Leicester had any chance of winning, and for several good reasons. Let me introduce three possible reasons why Leicester was not considered to be a candidate to win. I use these reasons to begin developing the discussion about small systems, such as soccer teams.

The first possible reason why Leicester was not considered to be a favorite to win the Premier League is that the team ended the previous season (2014–2015) ranked 14 out of 20, which means that it was ranked near the bottom of the league's table, with 65% of the teams ranked higher. While improvement from one season to the next is possible, climbing from near the bottom to the top of the league is less likely, unless the team has been miraculously improved, or its opponents have dramatically deteriorated.

The efforts required to climb from the bottom to the top are intuitively familiar to us but can be further emphasized through the *Elo rating* system.[3] Elo rating is a scoring methodology mostly used in sport. It is very clever in that when the system updates the score of a team or player after a match, it takes into account the prior *difference* between the ratings of the two teams or players. The odds of team G winning are calculated using the prior difference between the rating of team G (e.g. ranked at the top of the league) and the rating of its opponent, team L (e.g. ranked at the bottom of the league). The bigger the gap is, the less surprised we are when we observe the victory of team G.

We can further illustrate this idea using the context of boxers' ratings. Let's imagine a boxing match between two people: Mike Tyson ("The Baddest Man on the Planet") in his days of glory and the author of this book, who has never participated in a boxing match. Each fighter arrives at the match with his ranking in the table of boxers' current standings. Tyson is ranked at the top and the author, who has never participated in a boxing match, is

---

[2] https://www.espn.com/chalk/story/_/id/15447878/putting-leicester-city-5000-1-odds-perspective-other-long-shots-espn-chalk.

[3] https://fivethirtyeight.com/methodology/how-our-nfl-predictions-work.

by default ranked at the bottom. Beyond ranking there is reputation. Mike Tyson was known as one of the most formidable boxers and his reputation as "The Baddest Man on the Planet" was fully justified. The reputation of his opponent is academic only…. Therefore, the difference in ranking is such that no surprise would arise from observing Tyson winning the fight. Following such a win, the change in Tyson's Elo rating would be marginal, *regardless of the fact that he won.* The key point is not whether you win or not per se, but whom you beat or to whom you lose.

When we experience surprise, it is based on *expectations* that are grounded in *differences*, and this is an important fact taken into account by the Elo rating system. In general, it seems that our cognitive processes are relative, not in the misused post-modernist sense but in the sense that what motivates us, and for very good reason (Neuman, 2017), is differences rather than absolute values. As I will show later, differences do not just exist in our mind but are meaningfully grounded in the most basic physical aspect of our world, in which our mind—just like the minds of all other beings—resides.

If you have ever taken an introductory course in philosophy, then you have probably met the great Greek philosopher Plato. Plato believed in *ideas* or *forms*, which are the true, abstract, and transcendental objects underlying our messy reality. This is a very interesting idea. However, if we seek to understand our messy reality as it is experienced by real, messy individuals, then we must understand the *modus operandi* of real cognitive systems, and the fact that this *modus operandi* relies on differences rather than on absolute values or abstract ideas that exist in the transcendental and imagined realm of philosophers.

The same "relational logic" underlying Elo ratings is probably what also underlies the fact that, in real life, we usually have to choose between alternatives. A pragmatic American president doesn't ask himself whether, on a scale ranging from 0 to 10, Vladimir Putin should be scored 5.2 or 7.57, or how close he is to Plato's idea of the "Philosopher King," but whether Putin should be opposed, tolerated, or supported given the known alternatives. This is a real-world context of decision-making that is grounded in differences, expectations, and alternatives, and it makes a very interesting point.

Today, when we talk about surprise or uncertainty, we do so in terms of information theory and Shannon's measure of information entropy. For Shannon, information is a measure of surprise that is inversely associated with the probabilities of a *random variable X* existing with possible outcomes x_i. A random variable is just a variable whose outcomes correspond with some random process. For example, the two possible outcomes of an unbiased coin toss are heads and tails. The probabilities of observing either heads or tails are equal (i.e. half) and therefore the uncertainty associated with the outcome of tossing the coin is maximal. When an unbiased coin is tossed, we have the maximal level of ignorance about the outcome in the sense that we lack any *prior preference* for one outcome over the other. This is a situation of maximal uncertainty and therefore maximal information. My students in the course "Introduction to Cognitive Sciences" usually find it difficult to understand why this unbiased coin has maximum uncertainty *and* maximum information. It seems to them contradictory that uncertainty and information are somehow the same. Information, according to Shannon, is a function of surprise. In the case of the unbiased coin toss, we can have no preference for one outcome over the other. Therefore, we experience full surprise from observing the outcome. Tricky but interesting....

The Elo rating system proposes something different from Shannon. The "value" of a player (or a team) is a function of our expectations, and these expectations are not simply grounded in the probabilities associated with the outcomes of a random variable but in the *difference* between the two players (or teams). Again, this point may be considered meticulous, but it is not. In the real world, decisions are made in the context of differences (Neuman, 2017) and this context is a *difference that makes a difference* (Bateson, 1972/2000).

Now, let us return to the boxing match and imagine the unlikely scenario of the 52-plus-year-old amateur nerdy boxer from nowhere miraculously knocking out Tyson in the first round using a powerful combination of hooks and jabs. In this case, following the match, my rating would dramatically improve. Rating is supposed to reflect *virtue* (i.e. skill), but real virtue (i.e. skill displayed in a time of emergency), as described by Machiavelli (see Hiraide, 2014), is evident only when one plays against all odds, or at least dares to challenge Lady Fortuna, the lady of luck.

For natural systems, such as human beings, odds are always a matter of comparison situated in a dynamic context of decision-making rather than an abstract game of stable probabilities; knowing that Tyson had arrived at our match fueled by alcohol and drugs would have dramatically changed the odds even in such a clear-cut case where he was fighting an amateur. This is another example of the point that I have made before about deeds *in context*. In the real world, your virtue should be judged not by your wise and impressive speech but by your deeds only. The Elo rating system illustrates the known fact that the effort involved in climbing from the bottom to the top is *nonlinear* and intense. They are nonlinear in the sense that there is no fixed ratio between your efforts and your success and that the efforts of climbing to the top are much more intense than what might be expected by assuming linearity. This is trivial but must be mentioned.

The efforts mentioned above are grounded in reality where *observables*, such as teams' rankings, have a unique *distribution* indicating the *constraints* that form its shape. A distribution is just the way in which the values of a variable are spread. When you toss an unbiased coin, you will find that in the long run the proportions of heads to tails approximate 1—in 50% of the cases we get heads and in 50% of the cases we get tails. The famous *bell curve* is another form of distribution. It is the distribution university administrators like to impose on students' grades (goodness knows why). This distribution means that most students' grades are located near the average and that only a minority of the grades are either very high or very low.

Have you ever asked yourself how certain distributions are naturally formed? The answer is that some *constraints* channel the probability of occurrence of different potential outcomes (e.g. students' grades). For example, if you examine the distribution of wealth in some countries, then you will see that there is a very small number of very rich people who possess most of the wealth and a very long, thin tail of poor people. This form of *power-law distribution* is indicative of the *constraints* operating on the system. If wealth were distributed only according to the blind strategy of Lady Fortuna, the observed distribution would be much more homogeneous, meaning that the same proportions of people would be observed within the different segments of the socioeconomic ladder. Now think about the efforts required by a young person to climb up the social ladder in a society where a power-law distribution of wealth is observed. The efforts required to change your ranking in a positive direction are therefore grounded in the specific form of the distribution, which in its turn is indicative of the constraints operating on the system and constituted by the system through its internal dynamic. You don't have to be a neo-Marxist to understand how a society maintains constraints through which the wealth of the few is preserved and the equal opportunities of the majority to increase their wealth are limited.

The old Jewish sages invented a mechanism for correcting such a bias, called the *Jubilee*.[4] During the Jubilee, slaves were supposed to be released, debts forgiven, and lands that people had been forced to sell returned to them. The Jubilee is an impressive idea for two main reasons. First, it was proposed in the biblical days when equality of opportunity was not an accepted social–political idea, not to say the liberation of slaves. Second, the idea was that once every 50 years the cards were shuffled. The justification

---

[4] https://en.wikipedia.org/wiki/Jubilee_(biblical).

for the Jubilee was that shuffling the cards might somehow release the system from constraints and stagnation. It expresses the same idea as injecting noise or entropy into a system in order to release it from some kind of physical fixation. Through such liberating moves as the Jubilee, the old corrupt European nobility could not have survived. Shuffling the cards would have immediately removed some of the useless figures of this social class who lacked any clear virtue either by chance or as the unfortunate result of inbreeding. I mention the idea of the Jubilee and constraints because flexibility is one of the potential virtues that I attribute to a small system, a point that will be discussed at length in this book.

The above discussion explains why it is very unlikely that a soccer team will dramatically improve from one season to the next. The efforts and energy required from a soccer team to climb the ladder are nonlinearly correlated with the odds against it, which in turn are grounded in the team's prior ranking, which is supposed to reflect the team's virtue given the system and constraints in which the team operates. An underdog may surprisingly strike gold once, but consistently climbing the ladder against all the odds is truly a surprising process. As we all know, Lady Fortuna rarely supports those who lack skill—or, as Machiavelli said, "Fortuna also is unkind" (cited in Hiraide, 2014, p. 103).

In cases when we are surprised by an underdog winning against all the odds, we may conclude (1) that it is a rare and singular anomaly that has nothing to do with virtue or (2) that we have somehow failed to see the underdog's hidden virtue. Therefore, the odds against Leicester were probably grounded in expectations based on the recent history of the team's performance and the objective difficulties inherent in climbing from the bottom to the top of the league over the course of a single season. To repeat, these expectations were grounded in *differences* in ranking that were supposed to represent the real differences between the teams' virtue. Differences in ranking, though, are not differences in virtue. Virtue is a multidimensional and complex issue that we cannot directly observe. It is actually the name we give to the complex mechanism underlying an observed behavior (e.g. winning a match). Ranking, in contrast, is simple, explicit, and one dimensional. Differences in ranking and differences in virtue should not be confused, as we will learn soon. However, we should also understand that we constantly use simple substitutes for the real and interesting thing we truly seek to understand. Abstract and simple representations seem to be at the heart of human thinking. In predicting the odds of Leicester winning the Premier League, we have therefore made use of a substitute (i.e. the team's rank) and simple differences in ranking to learn about the hidden virtue of the team, and used this substitute to predict the rank of the team at the end of the season. As we lack simple access to the complexity of the world, a common approach is to use simple, informative clues that may be correlated with the event we seek to predict. There is more about this in the next section; however, before we move on, we must remember that a system operates under constraints and that we use constraints in order to understand and predict the behavior of a system. The ability of a system, such as a soccer team, to surprise us may therefore be grounded in its ability to somehow overcome the constraints imposed on it.

Prediction, Fools, and Babies

While all organisms perform some kind of prediction, as implied from their anticipatory behavior, the ability of human and nonhuman organisms to predict the future is no less than a miracle. Predicting the success of a soccer team based on its recent performance is an instance of this remarkable ability and so is the ability of a talented boxer to anticipate the next strike of his opponent in order to avoid it. Moreover, when a soccer team acts as a collective, it attempts to predict the next move of the opposing team, not by placing a bet but simply by adjusting accordingly. This is truly a remarkable ability, especially when it is expressed in real time. However, we should recall that being able to predict the future has always been a craft perceived with suspicion. For example, in one of the Talmudic tractates,[5] known in Aramaic as *Bava Batra* ("Middle Gate"), a Jewish scholar by the name of Rabbi Johanan said that since the destruction of the first Jewish Temple, prophecy (i.e. prediction) had been given only to fools and babies. This saying is sometimes used to describe those who (unsuccessfully) try to predict the future as fools. We occasionally see in the media "experts" predicting the direction of the financial market and whether certain stocks are heading up or down. Critically examining their predictions, you cannot avoid recalling Rabbi Johanan's saying. However, the meaning of the saying may be more delicate than implied in its common use; it is not that prophecy (or prediction in our secular context) is impossible. According to Rabbi Johanan, this ability does exist but was *exclusively* given to fools and babies. The question is why. Another Jewish scholar, known as the Maharal (1512–1609), provides an interesting explanation: the fool and the baby lack any predisposition and therefore they are not biased when listening to the word of God as it appears

---

[5] The Babylonian Talmud is the central text of Jewish rabbinic scholarship and it consists of 63 tractates written in around the 3rd to the sixth centuries CE.

in a prophetic vision. Prediction, as we learn from the Maharal, is thoroughly biased, but the meaning of "predisposition" or "bias" should be clarified. This is both the charm and the problem with some traditional sages. While they give us some insightful directions, it is not always clear where exactly they are heading.

To explain a possible meaning of bias, we may turn to the context of historical research, as history is a discipline where the past is the focus of research, and the question is whether we can learn something from the past or history in order to predict the future. History teaches us nothing, as argued by Hegel, because what we usually "learn" from history consists of general and abstract patterns that have little if any relevance to the tailor-made solutions required by real-world decision-making.[6] Learning from the past that a difficult economic situation might lead to dictatorship, such as in the case of Nazi Germany, is a possible but questionable lesson. The recent Great Recession in the USA has not led to a dictatorship and the horrible economic situation of people living in North Korea has not shaken the grasp of their tyrannical regime. The simple historical "lesson" taught about the relationship between economic situations and the emergence of dictatorships doesn't seem to be a lesson at all.

The bias mentioned by the Maharal may therefore be interpreted as an attempt to use general and abstract patterns in concrete and real-world decision-making. In this sense, the patterns are like a *Procrustean bed*, which twists reality to suit its own simple and rigid form. Forcing these general patterns on the real world, similarly to a Procrustean bed, inevitably results in leaving some things aside, and these things are probably those necessary to predict the future or make decisions in real time. Chopping off a tall person's legs is a possible way of fitting him into a short bed, but this would be a painful experience he should definitely avoid. The same may hold for prediction and for our ability to learn *useful* generalizations from the past.

As argued by Davies and Walker (2016), "the most powerful simplifications are abstract and generally hidden" (p. 2). Using these abstract simplifications is the hallmark of science, but are they applicable to the messy realm where we actually need tailor-made solutions in real time rather than abstract scientific simplifications? To repeat, the pitfall in prediction can be interpreted as the bias of using general and abstract patterns inapplicable to real-world decision-making. The lessons that we may therefore draw from the Jewish scholars are that (1) unless you want to use the advice of fools (and babies) it is better to avoid prediction and that (2) predispositions, in the

---

[6] See this talk by the historian Steve Mason: https://www.youtube.com/watch?v=MoHmSFWohg4.

sense of abstract simplifications, might distort the judgment of the normal grown adult and therefore bias in prediction is almost inevitable. The more general and far more important lesson concerns the way tailor-made solutions can be flexibly formed within some brute and general patterns or expectations that we learn through experience and reasoning.

We all understand that our ability to predict the future is limited and potentially biased, as evident from the surprising outbreak of the coronavirus in late 2019 (I started writing this book during the quarantine forced on us in 2020). But is it so surprising to observe the outbreak of a pandemic or is the real difficulty in predicting the *exact timing* of an outbreak? We seem to be limited in our ability to predict the exact timing of events, specifically those *remote* from our near future or those lacking a *cyclical* aspect (e.g. seasons). However, in some contexts, we are quite good at guessing (1) nearby future directions and (2) possible futures (e.g. the outbreak of a pandemic), and at using these general expectations to tailor good-enough ways of addressing challenges. By reading informative cues, we can tell whether we are heading in a certain direction, but precise prediction is limited. Our predictive ability and its limits are not merely psychological issues. Surprising as it may sound, they are deeply grounded in the physical nature of our existence, and it is highly important to contextualize our psychology and cognitive processes—those describing memory, inference, and so on—in the basic physical reality in which we all live. Notice that our discussion is contextualized within the case study of Leicester City FC and the odds against it. The logic leading to these odds is the same logic that Leicester could have used to surprise us and to refute our prediction—an important point that I would like you to keep in mind.

As creatures living in a world generally described by an *irreversible* process, as expressed by the arrow of time, we may learn abstract *patterns* (i.e. the ways things are related) and *order patterns* (i.e. what comes first and what comes next) and use them to monitor and regulate current situations and to anticipate future events. The irreversible nature of the world is deeply reflected in the mind of any living creature, although, hard as it may be to grasp, this perception has no simple grounding in the basic laws of physics.[7]

It must be emphasized, though, that the *irreversibility* of our experienced world is not merely a psychological phenomenon but an experience corresponding with a fundamental physical fact that our mind, like the minds of other beings, reflects. Recognizing this fact is critical to analyzing any decision problem. As argued by Peters and Gell-Mann (2016), "The typical decision

---

[7] See Richard Feynman's talk "The Distinction of Past and Future": https://www.youtube.com/watch?v=VU0mpPm9U-4&list=WL&index=28&t=280s.

problem only makes sense in the context of a notion of *irreversible time and dynamics*—we cannot go back in time after the gamble, and our future will be affected by the decisions we make today" (p. 4, my emphasis). According to this argument, there is no sense whatsoever to modeling a mind existing in a *reversible* world. In a world where the prey can return to the past and correct its decisions in order to avoid the predator, the predator itself can do the same, and the result is an infinite chase in time. Decision-making (i.e. our choices and actions) and expectations are deeply grounded in the *irreversible* nature of the world and the fact that we cannot simply return in time or reverse the arrow of time along the same path that has led us to this point.

When I was writing this chapter, my daughter Yaara mentioned to me that in *The Hobbit* (Tolkien, 1937/2012), Gollum plays a game of riddles with the hero, Bilbo, and in one of the riddles asks him to guess what is the thing that devours all other things. The answer is time. Time has been associated with an increase in *entropy*, an important concept that will be later discussed. From a basic experiential and commonsensical perspective, we perfectly understand why time is metaphorically described as devouring all things. As time unfolds, sooner or later all things are destined to be destroyed and there is no way for us to return them into being.[8] The first and most basic challenge facing all living things is therefore to maintain their existence against what my friend the immunologist and theoretical biologist Irun Cohen once described in a private conversation as the "wolf of entropy." Maintaining functional and structural integrity in the face of irreversibility is a challenge addressed by both an individual bacterium and a soccer team.

Irreversibility, though, has a Janus face. As mentioned, it reflects our experience that the more time unfolds, the closer we get to what we consider as a disordered state. I will not get into the philosophical question of whether order is in the eye of the beholder or consider the exact sense in which the concept of entropy reflects "disorder." At this point, we may just rely on the pre-theoretical observation that all living beings are destined at some point in their life to turn from a structurally and functionally collaborating system of molecules, cells, organs, and organisms into a set of molecules spread out in a potentially greater number of combinations than has ever been the case for those molecules before. When they approach such a state, they stop functioning as a structural and functional whole, and we see no correlation between the behavior of these units now and their behavior when they composed this whole.

---

[8] At least along the *same path* in which they were destroyed.

When Shakespeare's Hamlet observes the clown throwing up a skull, he says, "That skull had a tongue in it, and could sing once" (Shakespeare, 1987, 5.1). The irreversible nature of the world is simply and deeply reflected in Hamlet's observation. One may argue about the different senses of disorder, but Hamlet's observation represents its ultimate meaning. The tongue and the skull were once *structurally and functionally* connected and the fact that they are not connected anymore is what irreversibility is all about. Reversing this state is impossible no matter how much effort is invested. The skull and the tongue cannot return to work together and one doesn't require any background in thermodynamics in order to understand Hamlet's observation and its implied meaning. Therefore, the irreversible nature of the world is deeply associated with annihilation and the loss of *structural and functional integrity*. The measure of entropy (i.e. Shannon's information entropy) is only one possible and quite limited way of estimating the extent of the important issue, which is the loss of structural and functional integrity.

Small systems such as soccer teams and families are functional and adaptive. They have some coordination of their elements, and they maintain and manage their structural and functional integrity in order to adaptively achieve their goals. The family is a unit focused on caring for and nurturing a new generation. A soccer team is a system for scoring goals, avoiding goals, and entertaining an audience. This complexity should be kept in mind, even if I model it through approximate and simple measures and ideas such as entropy. Shannon himself wrote a short paper in which he complained that information theory had become a "scientific bandwagon" (Shannon, 1956). I will follow Shannon's advice and use the concept of entropy itself cautiously while allowing myself more freedom and less caution in using the concept for simple exposition and didactic purposes.

Irreversibility also means that strong *constraints* are imposed on the *order* of things, and these constraints are the source through which order is imposed on the behavior of a system, even for a limited period of time. When constraints exist, the observed system's *degrees of freedom* are limited in a *lawful manner*, and this information may be accessed by intelligent creatures, whose mind is coupled with the observed system. In a world where there is no difference between "before" and "after," where cause might precede effect and where metamorphoses from one state to another are symmetric, no information is available, no prediction is possible, and no control can be gained over any situation. It is hard to imagine the nature of cognitive processes existing outside the arrow of time. Therefore, the most basic scientific foundation of the human mind—and in fact of all living systems—is the concept of irreversibility, which is reflected in the adaptive and flexibly adjusting minds of all individuals and groups living on this planet. However, things are not so simple, as will be later discussed.

To the best of our understanding, the irreversibility of the world and the existence of constraints were never intended to be used for precise prediction by stock market "experts" and financial advisors. However, they are the foundations on which the minds[9] of all living systems are grounded. These ideas will be extensively elaborated throughout this book, but for now we should just understand that the odds against Leicester are grounded in the team's recent history and our ability as cogent creatures to rely on irreversibility to identify general patterns, such as the relative ranks of soccer teams and their predictive power.

Our expectations are grounded in differences and differences are grounded in the irreversible nature of the world and the constraints imposed on its functioning. In a reversible world, bookies could not have calculated the odds against Leicester. The first conclusion here is that the irreversible physical nature of the world is necessary in explaining our ability to (1) learn patterns, (2) predict (or more accurately "expect") future directions, (3) make reasonable decisions or guesses through heuristics, and (4) maintain our structural and functional integrity in order to achieve our goals.

Given the importance of irreversibility, we should also understand the importance of reversibility. If we cannot travel back in time, how do we learn what comes before or after? How can we differentiate between "before" and "after" or "this" and "that"? How is it possible to identify a "difference that makes a difference"?

We may later ask whether the same logic of irreversibility is what underlies the formation and behavior of small groups. I have no intention to argue that soccer teams are formed only to address the challenge of existing in a world governed by an irreversible dynamic. The existence of soccer in particular and sport in general may have multiple explanatory threads from the evolutionary to the cultural. However, the behavior of small social systems may be better understood by their correspondence with basic challenges facing each and every living creature. In a world where expectations are used to form tailor-made solutions, flexibly performing through collective activity may show some clear benefits, at least in certain goal-oriented contexts.

At this point, I would like to conclude the section with a speculation framed as a question. Is it possible that our joy in observing a successful soccer team tailoring a creative solution in order to score a goal may be the same

---

[9] I use the term "minds" in a general and abstract sense corresponding with the system responsible for all cognitive activities, from perception and memory to reasoning and prediction.

joy we experience when we observe our own mind at its best? The audience of tragedies in ancient Greece and the enthusiastic fans observing Leicester playing in Leicester City Stadium may both share the excitement of observing a struggle under uncertain conditions where a ready-made solution is never at hand for reaching a *catharsis*. Catharsis is built on tension and this tension is all about a *difference* and the joy of releasing the tension and closing a gap. Catharsis exists in soccer when tension is released as a team scores a goal. Catharsis also exists for a colony of bacteria closing a gap between them and a nurturing source of energy. Our fascination with the behavior of small social systems may be deeply rooted in a shared logic underlying the behavior of the observed systems—the basic logic of all minds, bacteria and soccer teams alike. However, given that we live in an irreversible realm where learning, adjusting, and adapting are inherent challenges, creative tailor-made solutions are always appreciated. If you understand this point, you may also understand that while the overwhelming odds against Leicester were perfectly rational when grounded in our "simplistic abstractions," it would never be possible to fully dismiss the possibility of them winning. In a fully predictable world, soccer would not be such fun.

In sum, we have learned that all organisms strive to learn and flexibly use patterns in an irreversible world, and that the low expectations of Leicester winning the Premier League were probably grounded in a process of prediction applied by soccer fans and bookies alike. In contrast with wise gamblers, soccer teams are less interested in prediction and more in playing soccer and scoring goals. Their behavior, though, has a unique aspect related to learning and using patterns.

On Scaling, Giants, and Dwarfs

In the previous section, we discussed the importance of irreversibility for understanding our ability to make sense of the world by learning patterns and making predictions. Irreversibility means that some order is imposed on the behavior of systems and that our ability to adjust and predict may somehow be associated with learning informative clues indicating the existence of patterns and order patterns. In this context, we may move on to the second possible reason why Leicester was not considered a favorite to win.

If you examine the summary table for the Premier League for the season prior to Leicester City Football Club's win (2014–2015), you may notice that the three top-ranked teams (Chelsea, Manchester City, and Arsenal) are all located in big cities: London and Manchester. And, if you examine the soccer clubs that didn't achieve enough points to stay in the Premier League, you will find that they come from smaller places: Hull, Burnley, and White City (the last of which is located in the area of Greater London but not in the city itself). While we are fascinated by stories about heroes winning against all the odds (see Gladwell, 2013), these encouraging stories are exciting first because they express our wishful thinking and second because they represent the exception rather than the rule. Leicester was no exception, at least as expressed by the gambling odds for the 2015–2016 season.

As can be learned from the summary table for the 2014–2015 season, a cruel reality governs the table, and this reality can be explained through the concept of *scaling*. A scaling law is a mathematical function associating two quantities. For example, when we plot the metabolic rate of an organism against its body size or the salary of people against the size of the city in which they live, we may observe an invariant and universal form.[10] Scaling has been studied and popularized by the physicist Geoffrey West (2017), who argues that a city's size matters and in a very measurable sense. The scaling law of cities is relevant to understanding the performance of soccer teams and for obvious reasons. First, the probability of recruiting soccer talent in big cities is likely to be higher simply because the population is bigger. Moreover, big cities provide opportunities that do not exist in small places (e.g. higher salaries) and therefore they can better attract soccer talent from smaller places. In their turn, the players contribute to the success of the city's soccer team in a kind of positive feedback loop that increases the attractiveness of the club to other young talent from small places. Even if you are not a soccer fun, you may be well aware of the fact that successful soccer teams do not usually come from very small cities, not to say villages. The chance of a talented soccer team somehow popping up in Oban, Scotland, or Urbino, Italy, is extremely small. While, to the best of my knowledge, bookies seem to have no specific interest in physics in general or in scaling laws in particular, their odds against Leicester City Football Club seemed to reflect a scaling law and its implied consequence: dwarfs don't usually win in battles against giants. Here we can see, *again*, how the physical reality is connected to our basic experience and understanding of the world in general and small social systems, such as soccer teams, in particular. My discussion of scaling therefore serves the same aim as the discussion of the previous section: to emphasize the importance of constraints for understanding the behavior of a small social system and the

---

[10] This universality should be examined with a grain of salt. Fitting a function to a set of observations always involves the danger of over-fitting. Testing the predictive power of a city's size against its team's ranking, for instance, may be better done using machine learning algorithms with a procedure known as *cross-validation*.

way a particular small system, such as Leicester City FC, may "free" itself from constraints and surprise us. However, at this point I would also like to highlight another important source of information that is grounded in our basic physical reality. To introduce this source, I first turn to the heuristics human beings naturally apply when reasoning.

The odds against Leicester may be grounded in a law of scaling that associates the rank of a soccer team with its city's size, but they probably stemmed from the successful heuristics people naturally apply when reasoning given their intuitive understanding of the scaling law. When asked which of two compared cities has the larger population, people can successfully rely on a discriminating cue, such as whether the city has a team playing in the top soccer league (Gigerenzer, 1997). The existence of a soccer team playing in the top league has no *causal relation* with the size of the city. A team playing in the top soccer league probably doesn't increase the size of its city, but it is reasonable to hypothesize that the size of the city is somehow *associated* with the success of the local soccer clubs. A scaling law may mathematically describe the function associating the size of the city with the success of the local soccer teams. However, the mechanism underlying the scaling law is somehow more difficult to understand as it involves a shift from *description* to *explanation*. While the description represents the association in mathematical terms that fit the empirical data, an explanation strives to identify the *causal* factors underlying the observed regularity. Such an explanatory mechanism may expose the way in which the size of the city results in the success of the local soccer team. Analogically to a metabolic network, one would probably not find a simple and direct mechanistic explanation associating the size of the city with the success of the soccer team but a complex network in action. However, despite the complexity of this network, we may recognize simple informative *cues* that may help us to guess which city is bigger and which soccer team is probably better.

Reasonable expectations usually don't require a full understanding and sometimes we may settle on using a few informative *clues*, correlations, and heuristics to model the observed system. These associations are the product of the constraints we have discussed before. Previously, I mentioned von Neumann, who said that science is about modeling rather than understanding and explaining. It is possible that to form reasonable expectations, we may just need to model the environment without meeting the high bar of understanding and explanation. We may just need differences (as illustrated by the Elo rating system) and similarities or associations (as illustrated by the scaling law), and these associations may be reduced to a few informative clues. What is true of science may be true of the mind, whether the mind of a bookie or the collective mind of a soccer team seeking to model and adjust to the behavior of the competing team.

Having a soccer team in the top league is indicative of a city's size as there is a scaling law associating the size of a city with the success of its local soccer team. This correlation or association means that some *constraints* are imposed on the behavior of the observed system. The meaning of *constraints* should be contrasted with the "anything goes" approach: it is not true that "anything goes." Some configurations or trajectories of the system are forbidden or less likely, and the "rule-based judges" that forbid the existence of certain configurations or trajectories are what we may call "constraints." This lawful behavior is expressed by the function associating the size of the city with the rank of its football team. In this context, the constraints are expressed by the fact that we see a recognizable relation between the two measurables (i.e., the size of the city and the rank of the team). Otherwise, we would not have observed a lawful pattern.

When constraints are themselves subject to constraints, we may learn a lot about the behavior of the system. For example, we may notice that the constraints imposed on the formation of patterns are constrained by time. Constraints reduce the entropy of a system or its degrees of freedom, and when we observe a system operating under some constraints, such as the constraints forced through *interaction* with another system, our uncertainty about the system's behavior is reduced and our ability to anticipate the system's behavior increases. To repeat, the natural world involves *interactions*. These interactions imply *constraints,* and these constraints entail associations that are expressed in the *reduction of uncertainty and the identification and learning of patterns through the heuristics of using informative clues*.

When observing a tennis ball bouncing along the ground and releasing energy through friction, we actually observe the interaction of systems (i.e., the ball and the ground) and the way the constrained behavior of the tennis ball expresses regularity *over time*. When we have played with a tennis ball for long enough, we intuitively understand that if it is left on its own, the bouncing ball will finally rest on the ground.

Imagine a child playing with a tennis ball in her room. She powerfully throws the ball to the floor and then immediately leaves the room. An hour or so later, she returns to the room to find that the ball is still bouncing.... This imagined scenario sounds like the opening of a horror movie and for good reason. Even a child who has never heard about the idea of *friction* intuitively understands that a bouncing ball loses energy to a point where it is motionless. A second source of information is therefore grounded in the interactive nature of the physical world and in our ability to learn informative clues by identifying interactions and their implied constraints and correlations. A mind interacting with the world is subject to constraints just like those imposed on the bouncing ball. Interactions impose constraints, and constraints imply correlations and the reduction of uncertainty.

The difference between a tennis ball falling to the ground from 7 m and another ball falling to the ground from 1 m is expressed in the height to which the balls bounce. The difference between the bounce heights is indicative of the heights from which the balls are released. We learn the association between the height from which a ball is thrown and the bounce height because there is a constraint on the bounce height, and this constraint is deeply grounded in the irreversible nature of the world. In a world where tennis balls randomly bounced in a way that had nothing to do with the height from which they had been thrown, ball games would look totally different.

Let us further elaborate on this example by assuming that a ball bounces to 60% of its original height. If it is thrown from 200 cm, it bounces to 120, 72, 43, 26 cm, and so on. We may follow the time-series of the heights and translate it into an order pattern of ranks. For instance, the three first values are 200, 120, and 72 cm. The highest score of the three is ranked "2," the medium score is ranked "1," and the lowest score as "0." Running a sliding window of size three on the time-series and breaking it into successive blocks of length 3, we notice that the time-series moves from order pattern {2, 1, 0} to order pattern {2, 1, 0} to order pattern {2, 1, 0}. The constraints imposed on the bouncing height through friction can be expressed by this order pattern, telling us that a *monotonic decreasing pattern* of the form {2, 1, 0} is always followed by exactly the same monotonic decreasing pattern *regardless* of the specific heights or the specificities of our rubber ball. We can start throwing the ball from 100 cm, we can use a more elastic ball that bounces to 70% of its height, and so on. In all of these cases, though, when *abstracting* the order pattern through relative ranking, we will still identify the *same general trajectory* of the system where the height of the bouncing monotonically decreases. For naïve scientists which is most of us—such a procedure may be enough to form reasonable expectations.

Now think about a football team. The interaction between the players (e.g., ball passes) imposes constraints and reduces the uncertainty within the team. Imagine a situation where there is exactly the same probability that each player in the team will pass the ball to any of the other players. In this context, the distribution of the ball passes is homogenous, meaning that the probabilities of a player passing the ball to any of the other players are the same. If we include the goalkeeper then the probability of each player passing the ball to any of the other 10 players is simply $p = 0.10$. This is a situation where the entropy of the ball passes is maximal. In this situation, predicting to whom the ball will be passed is impossible, communication between the players is defected, and the collective activity of the team cannot be orchestrated.

For example, we can imagine a team that includes only three players: the author of this book, Ronaldo, and Messi. When Neuman gets the ball, there is an even probability that he will pass the ball to Cristiano Ronaldo or Lionel Messi ($p = 0.50$). Both of them are perplexed as each of them faces full uncertainty. Who is going to get the ball? Should either of them approach me when I'm holding the ball, as he expects a pass? In this context, no preferences or expectations can be expressed, no prediction is possible, no communication is evident, and the uncertainty is full.

We now understand why constraints are so important for communication and the formation of order. Constraints formed through interactions are necessary for the formation of order and the possibility of communication, but constraints have a catch: they don't necessarily simplify our understanding but sometimes add complexity to possible solutions to a problem. As explained in the context of *combinatorial optimization*:

> The constraints reduce the size of the search space. At first glance this seems to facilitate the search for an optimal solution. The opposite is, however, frequently the case: Many optimization problems which can be solved efficiently on a computer without constraints, become extremely computer-time consuming if constraints are added. (Hartmann & Weigt, 2005, p. 2)

A simple humorous example may explain this. You are the head of a university department and are trying to organize your department's teaching schedule by matching classes (e.g., Introduction to Critical Theory of Educational Oppression), time slots (e.g., Sunday 10:00–12:00), and professors (e.g., Prof. Y. Y.). At first, it seems that you don't have a problem in finding a slot for each course. Although you have several courses in your department (e.g., Why Positivistic Research is Oppressive Research by Prof. H. P.), there seems to be no problem as you have several time slots in which you could easily locate any of the classes. However, here come the constraints. Prof. M. W. argues that he has difficulties in getting up in the morning and that his mind is not sharp enough before he drinks his morning coffee and has read *The New York Times*. He also requires some time to overcome his long weekend and to prepare for the next one, and therefore he is only willing to teach between Monday and Thursday, 12:00 to 14:00. Prof. H. P. argues that given her loaded schedule as a political and social activist she can be at the university

only between Monday and Wednesday and can teach only between 12:00 and 14:00. Similar circumstances apply to other members of the department. As you can see, constraints don't necessarily support the idea of order or an optimal solution, and this well-known fact is deeply associated with what is known as *the price of anarchy*: the degradation of the system, which is directly associated with the selfish behavior of its members. Selfish behavior may overload the system with constraints that increase the complexity of the solution up to the level where the problem cannot be solved. Having been the head of a university department for four years, I have observed the price of anarchy firsthand. As explained to me by a colleague, being the head of a university department is like being a cats' shepherd….

Theoretically, the freedom of the individual is maximal. However, no man is an island, and, whether we like it or not, as we live in widening social circles, we should be aware of the price of anarchy and the way in which it might impose impossible constraints on a system and its ability to function. Here we come to another important point about constraints.

Constraints are necessary for the formation of order, but we must be aware of the fact that they tend to accumulate in a nonlinear fashion. When there is a single player, he is of course theoretically free of any constraint. The idea of the Lone Ranger is the ultimate expression of freedom as he lives with minimal constraints. In the context of soccer, this freedom implies that no collective activity is possible as by definition soccer involves cooperation within a team and competition between two teams. When there are two players in a team, some constraints on the way they pass the ball to each other are added and these constraints accommodate as the number of players increases. At a certain point, there may be a kind of a *phase transition* where the system shifts to another regime and becomes *saturated* with no ability to change. As explained by Martin et al. (2001), "From a statistical mechanics perspective, a phase transition is nothing but the onset of non-trivial macroscopic (collective) behavior in a system composed of a large number of 'elements' that follow simple microscopic laws" (p. 3).

A phase transition in physics is quite different from that observed in small social systems. First, it concerns only large systems. Second, it results from slight modification of an order parameter, such as temperature. In the case of small social systems, we probably don't observe a "phase transition" as an abrupt change resulting from a slight modification of an order parameter. However, in both cases we observe a composite system shifting to a "nontrivial" behavior. I will use this sense of "phase transition" with regard to small social systems, which are the focus of our interest.

To gain an intuitive understanding of this non-trivial behavior (i.e., phase transition), think about how people may join a riot as individuals but turn into a faceless mob. Beyond a certain number of participants, the group starts functioning as a group rather than a group of *individuals*. At this point, adding more people to the situation would not change it. After it reaches its "mob phase," no individual can change the behavior of the mob from within. In such a case, and as in the *Ising model*,[11] some amount of disorder must be injected into the system in order to free its individuals from being aligned particles in a faceless mob. *An optimal level of disorder is therefore necessary for flexibility, change, control, and adaptation.* Please remember this important point.

When constraints are accommodated to a degree where the contribution of particles and their configurations has a minor influence on the behavior of the system as a whole, the system is frozen and cannot bootstrap itself out of its current organization. In such a case, the benefit of the small system is lost. *Ipso facto*, we may define small social systems as those where the system approaches a phase transition when scaled up beyond a limited number of particles and where the phase transition involves the loss of individuals' contribution to the functioning whole, whether a soccer team or a jazz trio. Please also keep this highly important point in mind.

In some contexts, constraints start to accommodate quite quickly and to a degree where no easy solution can be found. In fact, in the context of combinatorial problems such as graph coloring or Boolean satisfiability, it has been shown that there is an exponential function associating the system's degrees of freedom with the number of the components. The system's degrees of freedom exponentially *decay* as the number of components *increases*, and typically a sharp-threshold phenomenon occurs in the entropy, as it drops from positive to zero (Friedgut & Bourgain, 1999).[12] While this result comes from a limited and highly abstract theoretical scientific field, it has some correspondence in the realm of real small social systems as well.

Think about the scenario of preparing a dinner. When you plan to prepare a dinner for yourself and your wife, some constraints are naturally imposed, but what happens if you decide to invite your mother-in-law? Your mother-in-law is an orthodox Jew who insists on strict kosher food. She also loathes soy, which excludes many Asian dishes. Your father-in-law may be sensitive to

---

[11] https://www.quantamagazine.org/the-cartoon-picture-of-magnets-that-has-transformed-science-202 00624/.

[12] I thank my colleague Dr. D. Vilenchik for elaborating this point with me and for directing me to this reference.

gluten, which excludes all kinds of bread and pastries. Now add your brother-in-law, who is a vegetarian, or your sister-in-law, who can't stand garlic. Now the single optimized solution that you have been seeking for the dinner's menu has long gone. The reasonable solution is to increase the variety of dishes, but you would rather not put in such an intensive level of effort. After all, you were planning a nice family dinner rather than opening a restaurant...

Think about another example—the behavior of a couple and their free-time leisure activities. We will illustrate this example through a caricatured couple from the TV series *The Simpsons*. The father—Homer—may mainly enjoy spending his free time going to sports events and drinking beer (probably Budweiser) with his friends in the local bar. In contrast, his wife, Marge, enjoys staying at home, watching romantic comedies, and chatting with her two bitter sisters. Now, think about what happens when this couple interacts and seeks to enjoy their free time together. It is clear that both Homer and Marge will impose constraints on each other and that these constraints will shape their behavior as a couple; Homer has no interest in Marge's leisure activities and vice versa. Their interaction will entail constraints, but these constraints will not necessarily limit the "search space" of leisure activities up to a potentially frozen state; they may also generate complexity that enriches the behavior of the couple.

In some contexts, the constraints that emerge through interaction not only limit but also enrich our behavior, and this enrichment is probably the magic sauce that makes the behavior of small social systems so interesting and gives it a surprising aspect. As I previously stressed, the systems that we are discussing are quite different from those studied by physics, and their particles are not simple components that can be arranged in different combinations to find an optimal solution to a problem in a limited number of dimensions. The constraints definitely exist, but, when interacting under constraints, family members and football players communicate in a way that may push the context beyond that of solving combinatorial problems. In fact, the search space itself may be increased as a result of the interaction and the emerging new unit (e.g., a couple) may have unique properties and solutions that cannot be reduced to those of its building blocks. We will later discuss this important point.

The constraints also entail correlation between the behavior of Homer and Marge when they both operate within the same unit of interaction (i.e., as the Simpsons couple). This is because when they behave as a couple, they do the same things (e.g., going out for a romantic dinner), which are quite different from what each of them does as an individual or with others. That is, the interactions increase the correlation between the behaviors of the components and decrease the correlation between the behavior of the couple and the behaviors of each of its constituting units. To define a small social system, a *boundary* must be formed, and this boundary is formed through increased correlation between the particles when they interact within the new structural and functional unit (e.g. the couple).

In some cases, the constraints formed through interaction may even lead to novel behavior that doesn't characterize either Homer or Marge. Through interaction, constraints, and association, some small systems (such as the Simpsons couple) may naturally be led into creative and novel behavior that can be adaptively used. The form of creativity as expressed in the behavior of couples is not the same as we attribute to people like Leonardo da Vinci. It is a local form of creativity where the particles generate for themselves tailor-made and ad hoc solutions. The mutually imposed constraints force the Simpsons' system to produce novel solutions for cooperation. It is possible that we enjoy watching the Simpsons for the same reason we may be impressed by the behavior of other small systems, such as a soccer team, in action. Interactions form constraints that imply correlations but also a new level of organization with the potential for creativity. The novelties are not ready-to-wear approaches, but tailor-made solutions flexibly formed to support cooperation (within the system) and adaptation (to outside systems). Given the enormous differences between Homer and Marge, the Simpsons couple could not have survived as a couple without some creative solutions....

Now, let's return to the scaling law. The scaling law associating the size of a city with the success of its local football team may be used by human beings to produce reasonable expectations—or, better called, educated guesses. When considering the odds against Leicester, the bookies may have used several cues, including the city's population. As we can see, physics provides the foundational scientific justification for the odds against Leicester while cognitive research explains it in terms of the reasoning heuristics used by human beings. However, in this case too, a reasonable behavior is performed with no need for deep understanding. The gamblers betting against Leicester could have done so without being introduced to the idea of scaling or to the specific scaling law associating a city's size with the success of its local football team. The minds of bookies interacting with the world may have evolved to learn from the physical irreversible world and to identify correlations between systems by relying on the *mutual information* of cues. And it is not only the minds of bookies—the minds of every possible being on earth are designed to do the same. It is of no relevance whether you are a (neo-)Darwinist or an enthusiastic creationist. A mind that is not attuned to the world is of no use. This is why it is possible to model the behavior of plants by using the idea of *mutual information* (Zu et al., 2020). Surprising as it may sound, mind, in its general computational sense, is evident even among organisms that lack a central nervous system, such as plants. Plants maintain a complex network of interactions with insects and this collaborative activity cannot be modeled using a simple mechanistic model.

It seems that the most appropriate way to think about the mind of plants is that it exists *in between* the plant and the insect. This is an idea that in a general sense was proposed by Gregory Bateson (Bateson, 1972/2000; Harries-Jones, 1995, 2016). *Instead of thinking about the mind as a simple computational device exclusively located inside our skull, we may think about it as a process of computation carried in between particles that, when they work as a collective, form constraints through which ordered behavior and adaptive solutions are formed.*

Through understanding this complexity, we may be in a better position to understand that the world is not so simple as it may seem, and simple and brilliant abstractions, such as the idea of scaling proposed by West, are limited when we try to understand real-world decision-making—whether this is the decision-making of plants or soccer players. This important point is further elaborated in the next section. Please notice, however, that we follow the same path of explaining the odds against Leicester while, at the same time, explaining why the sources of the odds against Leicester can be used to explain why the team could have caused a surprise by winning against all odds.

Collective Dynamics and Surprise

A third reason for the odds against Leicester could have been the players composing the team. One may wonder whether beyond scaling and heuristics, it could have been that during this specific season, Leicester City Football Club managed to miraculously gather a group of talents who could have played in better teams for much better salaries and chose Leicester for other reasons, none of which I can easily think of. However, players such as Jamie Vardy, Riyad Mahrez, and Leonardo Ulloa did not fall into this category, and there is no reason to believe that the sum (i.e., aggregate) of Leicester's players was any better than the sums of those teams that were the favorites to win the league. Here comes the catch that we have already discussed. Small social systems, such as football teams, are characterized by emerging behavior, and this emerging behavior is sometimes hard to predict using the aggregate (or the simple sum) of the players' individual characteristics. Why? Because the "information" doesn't exist in the players but in between the players, as will be explained later.

Our mind is extremely limited in its ability to analyze collective behavior. Sometimes the behavior is additive, and the team's performance may be modeled as a simple sum (or an average) of its parts. In other cases, using an additive function might be misleading. This is probably what happened when the bookies evaluated Leicester. Indeed, the sum of the players' characteristics was not impressive enough to grant Leicester the status of a candidate to win the season. However, in some cases the behavior of the whole is non-additive and the particles' interactions form a surprising result on the macro level. This is clearly a potential benefit of a small social system. The magic sauce of such systems is their interactions and the way they may produce added value beyond that represented by the group's average performance. It is not the players' value but rather *the value of the players' interactions that gives the team* its Gestalt-like property of a whole that is different from the sum of its parts.

Generating unexpected results may have clear benefits to any system seeking flexibility and adaptability. A soccer team with behavior that is the sum of its particles' behavior would be totally boring to observe, not to say easy prey for its opponents. This is a highly important point. One of the potential benefits of a small system is its *ability to surprise* in real time and in more than one sense. We should notice that a phase transition of water into ice is an emerging and "surprising" phenomenon, but, once it is settled in a certain state (e.g., ice), the system doesn't surprise us anymore. Compare this behavior to that of a soccer team, where surprise potentially accompanies the team throughout *all of the match*, or even the behavior of a jazz trio, which keeps surprising us even as it improvises on known themes or pieces that we have heard before. This is the kind of surprise that I'm talking about, and I will elaborate the point further below.

In sum, the odds against Leicester may have resulted from (1) the team's performance in the previous season, (2) the clue associating the success of a team with its city's size, and (3) the "average virtue" of the players during this specific season. All of these clues are deeply grounded in processes that we as human beings, belonging in part to this physical reality, know how to use, at least to some extent. However, the most important lessons that we have learned through the above analysis are much more general. We have learned that small social systems face, like each of us, the challenge of living in an irreversible world, that they have the potential to express tailor-made solutions, that they involve interactions-based constraints that lead to associations and in some cases may lead to novelties through some kind of synergy between the particles, and that the magic sauce of their unexpected and surprising behavior may somehow result from their non-additive interactions and the fact that the collective mind of the system exists "in between" its particles. These important traits serve a major aim, which is to produce real-time flexible and adaptive behavior in a world where irreversibility imposes some constraints on our ability to learn and to practically use patterns, and where a delicate balance has to be maintained between learned patterns and the flexibility of tailor-made solutions. These ideas will be further elaborated in the following sections—but first to Leicester's surprising achievement.

Against all the odds, Leicester surprised almost everyone by beating the favorites and winning the Premier League.[13] While luck definitely plays a role in soccer, the sport is mainly a game of group talent—or, better called, *group virtue*—to include the unique and hardly mentioned talent of addressing the capricious nature of Lady Fortuna. This incredible virtue leaves little place for luck per se as the season unfolds, and Fortuna—the lady of luck—seems to align with the laws of probability. The lady of luck doesn't play football and Leicester's victory cannot be fully attributed to her grace.

The shocking achievement of Leicester is difficult if not impossible to understand and dismissing it as an "anomaly" doesn't explain what really happened or how this specific team won against all the odds. As wisely explained by Darwin in *The Origin of Species* (1859/2010), "chance" is just a term we use to express our *ignorance*. Saying that Leicester City Football Club won just by chance doesn't resolve our ignorance but confirms it in a circular manner.

It may be difficult to argue with the validity of the scaling laws identified by West, and we may reasonably agree that there is a scaling law associating a city's size with the success of its local football team. *Pre-theoretically*, and through our experience, we are well aware of the fact that bigger and wealthier football clubs have a clear a priori advantage, although this advantage is *statistical* only, leaving space for surprise and "anomalies" such as Leicester's win. The statistical nature of reality is such that by definition, it leaves enough freedom for surprise. However, the statistical nature of reality should not undermine our efforts to understand and model the surprising and emerging behavior of small social systems.

If there is a possible explanation for Leicester's success, then it will probably be found in the *collective dynamics* of the team and the way it formed a whole different from the sum of its parts. The club formed a whole that was

---

[13] It is estimated that the betting industry lost £20 million over Leicester's win. See https://www.bbc.com/news/uk-england-leicestershire-37037652.

much better than the simple sum of its parts (i.e., the players' performance as isolated particles). In contrast with its outcome (i.e., the victory of Leicester), this hidden dynamic is difficult to conceptualize, measure, and understand, and it definitely surprised the gamblers. In this context, relying on simple statistics about the soccer teams is not enough and cannot produce any significant or competitive advantage beyond what can be gained from common sense, intuition, and heuristics. As argued by a data analyst of football: "Statistics such as goals, shots, assists, and saves are easily understood and compared, but they are shallow representations of a complex, interconnected system" (Fletcher-Hill, 2014, p. 1). The same argument presented by Fletcher-Hill in the context of football is perfectly applicable to understanding the behavior of other small systems such as families.

In sum, what is really important about the Leicester City Football Club case study is that it exposes a deep lacuna in our understanding of small systems, such as soccer teams, and the way they form an emerging behavior through their components' interactions. Not only could we not have expected the exception of Leicester's victory—a "failure" perfectly explained by the statistical nature of reality—but we cannot even retrospectively explain the unique dynamic of the team even when we know that it has already won the season. What is the "difference that makes a difference" between the team that gets to the top of the league and those that gain second and third place? Is it luck only? Are all happy families alike and is each unhappy family unhappy in its unique way, as proposed by Tolstoy at the opening of his monumental novel? What is the unique signature of successful start-up companies? In sum, the surprise that we experience when observing Leicester's Cinderella story should increase our awareness of how little we really understand the dynamics of small systems, from soccer teams to families.

I have further argued that small systems present an important aspect of our nature, which is the ability to use tailor-made solutions to flexibly address the challenge of decision-making in an uncertain reality. This logic underlying the behavior of small systems is probably the reason why we are excited by both family and football dramas. Regardless of all simplistic abstractions and their appealing beauty and potential predictive power, the real thrill is in observing the real-world tailor-made solutions as formed in vivo. In this context, it seems that all successful soccer teams are alike in their ability to apply learned patterns of behavior. However, each successful team is successful in its own way, as it is able to craft the tailor-made solutions necessary to beat its opponent, which is also seeking the silver bullet needed to win the match. In a world governed by both competition and collaboration, this tension between ready-made and tailor-made solutions is inevitable. Now, the real question is how to better model and understand small systems. To try to answer this question, we may move on to the next section.

How Complex Can a Group of "Hooligans" Kicking a Ball Be?

Physics has made impressive strides in modeling populations of particles, such as gas particles, and the inevitable question is whether we can use this knowledge to model the behavior of *small* systems such as soccer teams. A soccer team, just like other small systems, does not fall under the rubric of the systems modeled by thermodynamics and statistical mechanics. First, statistical mechanics deals with systems that have huge numbers of particles and looks at how those systems are described by macro-scale properties (Poirier, 2014). As the micro level of such systems is impossible to describe given its numerous particles and their potential combinations, the interest of thermodynamics is in the macro-level properties of matter, as measured by a few variables: temperature, pressure, volume, and number of particles. However, in the context of small social systems, we do not have any problem with the number of the particles. We are dealing with *small* systems, which by definition have only a few particles: football teams, families, boards of high-tech companies, jazz trios, and so on. The difference between the micro and the macro levels seems to be irrelevant to the study of small systems and averaging the behavior of the particles gives us no interesting information beyond what may already be known by the naïve observer. Knowing that on average successful soccer teams hold the ball for a longer time than unsuccessful teams is of no great scientific or practical value.

In this context of averages, it must also be noted that the averages describing the behavior of a small system can be of two different types: the average of the *ensemble* in a time-slice and the average of a *single particle* over time. These averages do not necessarily converge, and in such cases, we say that the system is *non-ergodic*.[14] Small systems such as soccer teams seem to be non-ergodic, and modeling non-ergodic systems assuming ergodicity may involve deep flaws and wrong conclusions (Molenaar, 2013; Peters & Gell-Mann, 2016; Taleb, 2020). This point will be elaborated in Chap. 3 when we discuss the potential benefits of small systems and why playing

---

[14] A friendly explanation of this point appears at https://taylorpearson.me/ergodicity.

Russian roulette is not such a good idea. In the meantime, think about a small system whose behavior cannot be averaged in any meaningful way. Isn't this a good way to behave in a world mainly characterized by non-ergodic processes? When answering this question, compare the non-averaged behavior of a small system with the averaged behavior of some kind of population. Can you imagine a situation where a crowd's wisdom turns into stupidity as a result of an averaged behavior?

There are several additional differences between the systems modeled by thermodynamics and statistical mechanics, and small systems. In statistical mechanics, individual particles have infinitely marginal impacts on the behavior of the whole, on the macro scale. However, in a soccer team, a great player can make a difference that makes a difference. Can you imagine the success of Barcelona without the unique contribution of Lionel Messi? Or the success of Portugal in the 2018 FIFA World Cup without the contribution of Cristiano Ronaldo?

In addition, small systems present differential contributions of their various partitions. The "chemistry" between two football players (Bransen & Van Haaren, 2020) may uniquely characterize the pair and significantly contribute to the performance of the whole team. Other combinations may lack chemistry and may lead to additive behavior or even sub-additive behavior. This is observed both in soccer teams and in families. We observe couples that present "chemistry" but the recipe for this chemistry is far from simple, as is well known to every intelligent family therapist.

Not only are the systems we are dealing with too small to adopt the perspective of statistical mechanics, but also the "particles" composing them have well-defined identities and make differential contributions to the behavior of the whole, as they communicate with each other in a way that goes beyond mere interaction. When two billiard balls collide, each changes the direction of the other. When two human beings form a couple, they form a new and *unique* structural unit that is a product of their *repeated interactions*. The formation of such a novel structure has no equivalence in the world of the billiard balls.

In addition, in living systems direct interactions cannot be trivially scaled up to large populations and therefore we observe the use of signs, whether in human communication or in *stigmergy*, where interactions are mediated by traces left in the environment (as in ant communication). As explained by Davies and Walker (2016), "When there is substantial information transfer that is *not mediated* by a direct causal interaction, it is a sign of entrainment, or cooperative behavior, where collective modes of the system dominate over individual modes" (p. 6, my emphasis). When we observe a population of gas particles, no collective behavior is evident in the sense observed in social systems, but just averaged behavior. When we observe a swarm of ants communicating through pheromones, we observe information transfer (e.g., "here is a drop of sweet ice cream!") that does not necessarily use direct interaction. In small systems, we observe an interesting combination of direct and symbolic interactions that exists on a small scale. A couple having a verbal interaction or two soccer players passing a ball are using both direct and symbolic interactions. This mixture of direct and symbolic interactions is important as it combines the benefits of local and direct interactions with the benefits of general and indirect communication through signs. It also forms a watershed differentiating large systems (such as ant colonies) from small social systems (such as soccer teams or jazz trios). Swarm intelligence and "jazz intelligence" are probably different, and in this book, we are dealing with the latter.

Small systems also interact with other systems, and therefore they are adaptive. A "swarm" of gas particles has no interest in others, but a football match has no sense outside the context of competition between teams and a family has no meaning outside the context of the community of which it is a part. In psychology, we have for too long been indoctrinated to believe that we can model a particle by isolating it in a laboratory tube, in vitro. The overwhelming majority of my students who gained their bachelor's degree in psychology (just as I did many years ago) have never heard of Russell's paradox or Gregory Bateson's seminal work. They have also never been fully exposed to the *qualitative* differences between the different levels of analysis of a system, or the way in which the behavior of a lower level of a system is embedded within higher structural configurations. Individuals behave differently from small groups, which in turn behave differently from big groups and populations.

The emerging behavior of small systems is deeply grounded in the interactions between their members and in the interactions with other systems, and these interactions cannot be modeled without considering a collective dynamic involving information and communication at different times and organizational scales of the system. Interaction is one of the major concepts to be discussed in this book, and I will try to explain it in both a simple and a rigorous manner. The emerging behavior of a system is observed through interactions, and these interactions define the system's boundaries and its distinctive behavior. Statistical mechanics was not originally designed to model such systems and therefore one may suggest that the modeling of small systems belongs to the realm of psychology and specifically to the sub-field of *social psychology*, rather than to the realm of physics. I critically examine this suggestion in the next section.

The Psychologist and the Bookie

Psychology seems a natural choice when seeking ways to understand and model small social systems such as football teams. In contrast with physics, psychology's *raison d'être* is the study of the individual, the particle, rather than the population. Therefore, it is reasonable to expect psychology to provide us with some directions about the way a small number of particles work together, each being influenced by the social context in which it exists. However, modern experimental psychology has almost totally neglected the study of the individual and preferred a *nomothetic* rather than an *idiographic* approach, expressed in the analysis of groups of people and macro-level variables. Moreover, confusion between the individual and the group levels of analysis has led to some serious methodological flaws and invalid results, as exposed, for example, by the psychologist Peter Molenaar (Molenaar & Campbell, 2009). In many cases, these flaws result from the violation of ergodicity, but the basic fallacy is the inability to understand that small systems deserve a different approach from the one that has dominated psychology for years.

In sum, the study of the individual has been neglected. The price of focusing on group *averages* is that experimental psychology has failed to understand not only the uniqueness of the individual (as done by great novelists, for instance) but also the dynamics of small social systems of individuals and their emerging behavior. As such, psychology has failed to scientifically model and understand the individual as well as to model and understand the emerging behavior of small systems. This point requires no further explanation but just a simple illustration.

A discipline striving to be "science" but that cannot provide workable insights to enable modeling of real-world situations, should critically examine itself. Mathematics has no pretension to help us understand the world in any practical manner and nor do poetry and literature. However, paradoxically, through having no obligation to reality, they are highly successful in explaining it (see Neuman, 2020). However, scientific psychology is neither mathematics nor literature. Psychology that cannot explain (i.e., predict) the *real* behavior of *real* human beings is of no value. And now for the illustration. The sports gambling industry, whether you like it or not, deals with huge amounts of money and with predictions in the *real world* of sport. The real world, complex and obscure as it may be, is the only benchmark on which an empirical science can rely and the only playground for empirically examining the validity of scientific models. Better understanding the behavior of groups, such as soccer teams, could make a huge difference in predicting the outcomes of matches and make a lot of money. I am not an ideological capitalist and have never gambled, but one possible reason to understand small systems such as football teams may be the ability to systematically predict their success in a way that leads to a significant improvement over some gold standard or a relevant benchmark (e.g., the layman's prediction). This is a context where a *real* criterion for testing our understanding or predictions exists.

To the best of my knowledge, the sports gambling industry has never successfully and consistently used experimental social psychology in order to improve the forecasting of soccer tournaments. In a deep sense, this case is highly similar to one reported in the context of fintech (i.e., financial technology), where it is said (Zuckerman, 2019) that the mathematician Jim Simons avoided recruiting economists for his highly successful companies for what are probably *exactly* the same reasons that social psychologists are not recruited to the sports gambling industry.[15] The real world is a nasty place, dashing our fantasies again and again against the solid rock of reality. As the successful baseball player Yogi Berra is popularly attributed to have said, "In theory there is no difference between theory and practice – in practice there is." It is of no surprise that this insightful observation was perhaps produced by a person who was deeply immersed in sport. In theory, cognitive dissonance between what you believe in and what you do should produce stress, which in turn may lead to a change in attitude. But, as has been wisely and critically argued by the psychologist Billig (1987), this theory is refuted simply by observing the realm of politics, where it is common to observe inconsistency between what is (allegedly) believed and what is actually done. People are extremely talented at ignoring inconsistent evidence, especially if it contradicts their basic motivations, and therefore they have troubled relations with reality. As explained by the writer Upton Sinclair, "It is difficult to get someone to understand something if their paycheck depends on their not understanding it" (cited in Wainer, 2016, p. 1). Through reading Sinclair's wise observation, I better understood why academic psychologists are so pissed off at being criticized for the lack of scientific rigor and real-world relevance of their profession. Using a snowclone of a famous American

---

[15] A similar story is told about a famous poker player who established a successful real-estate company. He refused to hire employees with a background in finance (Komikova, 2020).

phrase, we can explain psychology's incomprehensible resistance to relevant criticism by saying: "It's paycheck, stupid."

Returning to soccer, we may wonder why social psychologists, who should be the ultimate experts in the social behavior of human beings, are not recruited to improve the predictions of gambling companies. After all, a soccer team seems to be a simpler case than a family or the management of a successful high-tech company. In fact, I have even noticed that some academics express deep scorn when talking about sport in general and soccer in particular. As once said by a famous Israeli philosopher (i.e., the late Yeshayahu Leibowitz), soccer is just a group of 22 hooligans kicking a ball.[16] So how difficult can it be to predict the outcome of a soccer match when you are an "expert" in human beings and all you have to observe is two teams of "hooligans" kicking a ball? Can social psychology develop a model that significantly improves the predictions of laymen and bookies alike?

The answer is evident from the non-existence of social psychology in the sports gambling industry. The reader is invited to challenge me by trying to recall the last time he or she saw a hiring advertisement for the sports gambling industry where social psychologists were invited to apply for a well-paid job predicting the outcome of soccer matches. Data scientists—sure; physicists, who are experts in quantitatively modeling complex systems—probably; but not social psychologists. Small social systems, such as soccer teams, are much more difficult to model and scientifically understand than we might imagine, and social psychology doesn't give us any constructive direction for studying them. At this point, it seems that we have to return to physics in order to find at least some direction for studying small social systems. Surprisingly, though, I will show how basic foundational ideas in physics are associated with basic cognitive processes and how the association between physics and cognition may point to a constructive way of studying small social systems. To address this challenge, we should first delve into some foundational issues in physics and consider how several foundational *concepts* may help us to approach the modeling of small systems.

Summary

- Leicester City Football Club was not a likely candidate to win the 2015–2016 Premier League season.

---

[16] https://en.wikiquote.org/wiki/Yeshayahu_Leibowitz.

- A small social system operates under constraints and we use constraints in order to understand and predict the behavior of the system.
- The ability of a system, such as a soccer team, to surprise us may therefore be grounded in its ability to somehow overcome the constraints imposed on it.
- The irreversibility of the world is crucial for learning patterns and making predictions.
- The irreversible nature of the world is deeply associated with annihilation and the loss of structural and functional integrity.
- Maintaining functional and structural integrity in the face of irreversibility is a challenge addressed by both individuals and small social systems.
- Constraints are necessary for the formation of order, but we must be aware of the fact that they tend to accumulate in a nonlinear fashion.
- An optimal level of disorder is therefore necessary for flexibility, change, control, and adaptation.

3

How to Understand Small Systems Through a Little Demon, a Drill Sergeant, and Indiana Jones

In this chapter, we learn about a little demon that has long challenged physics, gain some insights into irreversibility through an army sergeant, understand how memory is crucial to forming abstract patterns, and see how information is formed in between the particles of a system. Moreover, we learn why Nazis are wrong in their understanding of entropy, how Indiana Jones is associated with the law of requisite variety, and how tailor-made solutions are formed by small social systems.

Maxwell's Demon

The small systems discussed in the preceding chapters may be described as existing on the *mesoscopic* level of analysis. In the study of networks, the microscopic level concerns the properties of single nodes while the mesoscopic level concerns the properties shared by *groups* of nodes (Reichardt et al., 2011). We can therefore distinguish between three levels of analysis: the microscopic, dealing with the properties of individual "particles"; the mesoscopic, dealing with groups of individuals as defined through their interactions; and the macroscopic, dealing with macro-states and macro-variables of the population of individuals. Nothing has been said yet about time, which is a crucial factor in modeling small social systems. Small systems are not crystals. They are systems that exist in time and avoiding the study of their dynamic is impossible if we truly seek to understand their behavior.

© The Author(s), under exclusive license to Springer Nature
Switzerland AG 2021
Y. Neuman, *How Small Social Systems Work*, The Frontiers Collection,
https://doi.org/10.1007/978-3-030-82238-5_3

The lacuna in understanding *small social mesoscopic* systems invites the question of whether such systems, specifically in the social context, may be *scientifically* modeled at all. The current book tries to positively affirms this question and attempts to provide some basic scientific guidelines on modeling and understanding small social systems. The basic tenet of the book is that some foundational concepts can help us to understand small systems. By saying that, I don't conform to a simplistic reductionist approach arguing that small social systems can be modeled like other physical systems—a position rejected in the preceding chapters. Small social systems have their own unique signature, being at the same time small, cognitive, and social. In fact, they constitute a *unique category* by being *both* cognitive and social. They are cognitive in the sense that they are composed of "particles," where each is an autonomous computational unit representing the environment and operating on these representations and in and on the environment. Each player in a soccer team performs necessary computations in order to adjust his behavior to that of others. Small systems are also social in the most basic sense that through the interaction of the system's particles, the collective dynamics give raise to an emerging behavior through which the system is observed as a functionally and organizationally distinct whole, different from the sum of its parts. Therefore, in studying small systems the cognitive and the social dimensions are inseparable.

So why should we turn to physics in order to study small systems that are both cognitive and social? Shouldn't we return after all to psychology or sociology in order to find the answers? To answer this question, we should recall a prominent, albeit imaginary, figure in the history of physics, which is Maxwell's demon (Leff & Rex, 2014). Through this thought experiment, we may elaborate the foundational concept of irreversibility, understand how it is connected to computational aspects of the individual "particle," and see how this is all connected to the way small systems work in a world where both abstract patterns and tailor-made solutions are required. My use of this thought experiment aims to serve one major aim, which is to show how the social–cognitive aspect may help us to understand irreversibility and the behavior of small systems. The thesis and discussion developed in this chapter are presented in a nonlinear fashion where repetition is evident. All the threads, however, will converge at the end of the chapter, hopefully justifying the reader's patience.

Maxwell's demon is a thought experiment that aims to test the possibility of *violating the second law of thermodynamics*. Without going into too much detail at this stage, let us just recall that the second law of thermodynamics suggests that the *entropy* of a system increases under any "spontaneous irreversible thermodynamic change" (Poirier, 2014, p. 99). Simply and more generally stated, the law suggests that if you observe a closed system undergoing a spontaneous change, then this change is probably toward a state where a less "organized" configuration will be observed.

In this context, Maxwell's thought experiment involves a tiny creature (i.e., the demon) controlling a door separating two chambers containing gas molecules. The demon observes the molecules and as an intelligent creature knows how to differentiate between fast- and slow-moving molecules with no effort at all. Therefore, the demon can open the door at such times to allow fast-moving molecules to concentrate on one side and slow-moving molecules to concentrate on the other. The door is frictionless, and no effort is required to open or close it. In this way, one chamber becomes hot while the other becomes cold. See Fig. 3.1, which illustrates this idea. The slow-moving molecules are marked with crosses.

Now, the second law of thermodynamics suggests that the entropy of an *isolated* system can *never* decrease over time. If there is a *spontaneous* change in the system, then it is a change toward an *increase* in entropy, where entropy may generally be considered as a measure indicating the extent to which the particles are evenly distributed. In other words, "spontaneous processes have a *preferred direction*" (Lutz & Ciliberto, 2015, p. 31, my emphasis) and this direction is toward an increase in entropy. Here I would like to explain Shannon's measure of *information entropy* in order to better explain the above figure.

For Shannon, information is all about surprise and surprise in grounded in probability or the degree of belief we attribute to a proposition. For example, let's imagine that you are a big fan of Leicester City Football Club and before the season begins, you are sitting in your favorite local pub, named *The Dame and Turkey*, drinking beer and asking 10 of your friends which of them believes that Leicester is going to win the season. As expected, you

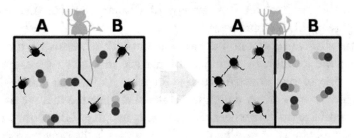

Fig. 3.1 Maxwell's demon. *Source* Wikipedia

are the only one who holds this strange belief—all the other people sitting with you hold the opposite opinion: there is no way Leicester will win the season. As the ratio of those believing Leicester will win to those who hold the opposite opinion is 1–10 (i.e. 0.1), the expected surprise if Leicester wins the league can be expressed as 1 divided by the probability 0.1 (i.e. 10). The fewer people who believe that Leicester will win, the greater the expected surprise if the team wins, and vice versa. For example, if 9 out of 10 people believe that Leicester will win, then the surprise of observing a win is relatively small (i.e. 1.11). After all, if 9 out of 10 people believed that it would win and the team won, then there would be no real surprise.

Now think about tossing an unbiased coin, where the probability of a heads result is equal to the probability of a tails result. In this case, our ignorance of the outcome is full because the equal probabilities cannot lead to any expectations beyond those formed by *full ignorance*. When observing the outcome of tossing a coin, you are fully surprised in the sense that nothing has led you to expect the outcome—or, better stated, nothing has led you in advance to form preferences for one outcome over the other. Following this line of reasoning, Shannon's *information entropy* is defined as:

$$H(X) = \sum P(x_i) \log \frac{1}{p(x_i)}$$

where the sign $\sum$ (i.e. sigma) asks you to sum the probability of x_i (e.g. heads) multiplied by the log base 2 transformation of 1 divided by the probability x_i, over all instances of x_i (e.g. heads and tails in the case of coin tossing). Therefore, H measures the *average surprise* existing in a *random variable*. For example, the information entropy of an unbiased coin is:

$$p(0.5) \times \log \frac{1}{0.5} + p(0.5) \times \log \frac{1}{0.5} = 1$$

This is the maximal level of entropy of a binary variable (i.e. a variable with only two possible outcomes).

We can also think about information entropy in terms of the process of finding the location of an object or particle in a given *phase space*. For example, one of our domestic cats—Gin-John, (known as Jake)—likes to sleep in the drawers of my daughter's closet. Let's assume that we have eight drawers and have no clue where the old cat is napping. Given our full ignorance, the probability of finding the cat in a drawer is the same: $1/8 = 0.125$. The corresponding Shannon entropy is 3. What does this mean? Think about playing an equivalent to the "20 questions" game, where one person imagines an item and another asks up to 20 questions to which the answer can only be either "yes" or "no," with the aim of identifying the item. To find the cat, you can ask one binary question at a time. You can begin by asking a question that splits your "solution space" in half. For instance, you can ask whether the cat is hiding in the bottom four drawers. Let's imagine that you get a "yes" answer, which removes the possibility that the cat is hiding in any of the top four drawers. Next you ask whether the cat is hiding in the top two drawers (of the bottom four). You get a "no" answer, and again your remaining solution space is reduced by half. Finally, you look at the two remaining drawers and ask whether the cat is hiding in the top drawer. After you receive the answer, you can locate the place where the cat is hiding, and you have done so with only three questions. This means the Shannon entropy is 3. "In such an approach, entropy equals just the number of YES/NO questions needed to locate the selected particle" (Wilk & Włodarczyk, 2008, p. 4810).

In the figure above, you can see that before the demon moves into action, the fast-moving molecules are distributed in the box. The distribution is fairly *symmetric*. In terms of entropy, if the distribution is even, then the entropy is maximal. After the demon has completed its intervention, the fast-moving molecules are located only in one chamber and now the distribution is *asymmetric*, and the entropy has been decreased to the minimum. Order has been formed.

Now that we have introduced the experiment and Shannon's information entropy, we may return to discuss the second law of thermodynamics. This law describes the arrow of time in terms of *energy* available to do some work (i.e. entropy) (Hemmo & Shenker, 2012). Here the concept of entropy is used in its original thermodynamic sense of energy available to do some work, but in a short time the connection with Shannon's entropy will become clear. The arrow of time "says that equilibrium states in which energy is less exploitable as work occur *later* than equilibrium states in which more energy is exploitable as work" (Hemmo & Shenker, 2012, p. 34, my emphasis). Please notice that, for an observer to recognize that an equilibrium state with less exploitable energy occurs *later*, he must have the ability to grasp the arrow of time, which is circularly defined through the order pattern of equilibrium states. This self-referential process seems to characterize the behavior of all living systems. However, this is not a book on philosophy of physics or biology, and my minor comment aims only to emphasize the inevitable difficulties facing us in trying to understand foundational concepts and the way they are relevant to studying the minds of living creatures.

At this point, we may better understand the above experiment in terms of *information entropy*. When the particles' distribution is even, and fast-

and slow-moving molecules homogenously populate the whole box, no work can be done. However, after the little demon has intervened, the particles' distribution has changed and now the fast-moving molecules are in the right chamber while the slow-moving molecules are located in the left chamber. When the small door separating the two chambers is reopened, the fast-moving molecules will flow out of it and this flow may be used to do some work, such as feeding a hungry steam engine. By doing this effortlessly, the demon seems to contradict the second law, which stipulates the order in which various levels of available energy appear to the observer.

Beyond the second law, we may interestingly understand that in order to do something in this world, there must exist some kind of demon representing a cognitive process that is necessary to identify available energy and use it. This energy can be expressed in terms of information; therefore, underlying every living system, we should find an information-processing device. More fundamentally, at the bottom of this "device" is the ability to identify *differences*, modeled as different entropy states, and to sort out "this" from "that." In a world where everything is the same for a given observer, such as our demon, nothing can really be done.

Identifying *differences* is crucial for all living beings. We are all fed by energy and identifying differences in entropy is necessary for our survival. By intervening at the *molecular* level of analysis, the demon forms a situation in which there is a difference between the two chambers' distributions of molecules, and where the entropy of the system on the macro level has been decreased. It is argued that "Maxwell's demon revealed the relationship between entropy and information for the first time, demonstrating that, by using information, one can relax the restrictions imposed by the second law on the energy exchanged between a system and its surroundings" (Parrondo et al., 2015, p. 131). This is an interesting argument although it involves some deep perplexities. Entropy is a measure that describes a system on the macro level of analysis, and it is actually a measure of uncertainty. Information, whatever it is, is something that the demon obtains on the micro level of analysis as he measures the speed of each and every molecule. How are these two related? In addition, the context is a closed system while living systems are open systems that exchange energy and matter with the environment. Is the little demon a part of the closed system or an outsider observing and intervening in it?

From a "practical" modeling perspective, we can understand how the demon actually converts information into energy. By being able to distinguish between fast- and slow-moving molecules, the demon can sort them into the two chambers to change the entropy of the system. By making it a low-entropy system, the demon can open the door and through the flow of molecules produce energy by using simple machinery, similarly to the way a steam engine works. The demon is therefore first and foremost a *cognitive* machine that, by converting information into energy, may sustain its own existence (for an interesting discussion on cycles of information and energy, see Alemi, 2020). Here you can see the connection between irreversibility and the mind; any system is destined to disintegrate. However, the temporary existence of a structural and functional whole, like you and me, is possible only through the existence of a "mind" through which information is processed and converted into energy, supporting the existence of the whole being out of equilibrium. The demon thought experiment was originally created to question the validity of the second law. In retrospect, it may also be used to discuss the relation between mind and life. Let me explain this point.

In a closed system, the *energy* available to do some work cannot spontaneously increase. While our mind is an open system, we know how to recognize differences in entropy because if we could not do this, we could not identify and use various sources of energy. To identify the energy available to do some work, we must first identify *differences*, and this cognitive computational work must be fueled by available energy. As explained by Alemi (2020), living structures are characterized by cycles of information and energy; we use information to gain energy and we must gain energy to gain information.

Cycles of energy use to identify available energy are at the heart of all organisms (Randall et al., 2002), and the challenge of identifying available energy has a clear cognitive dimension. My point is that our own existence far from equilibrium is cognitive through and through, given the continuous need to identify available energy as expressed in differences of entropy. Cognitive processes seem to have a foundational status no less important than those described by physics. The mind is foundational for the understanding of our existence no less than foundational concepts in physics, such as energy. Please keep this important point in mind as it frames the foundational idea of reversibility in the context of cognitive processes.

I have mentioned the idea of *difference* for a very specific reason. The most basic conceptual unit of information is a difference or a *distinction* (see Rashevsky, 1955), and one of the most basic distinctions is between high- and low-entropy states indicating available energy. We should recall an important point discussed by Bateson which is that a difference is an abstract thing. A difference can be expressed in matter, such as in the case of a mark left on a stone, but for a cognitive system it is an abstract thing as it is a "representation" existing in the system's mind. At this point, you may realize something that rather surprised me, which is that if a "Maxwellian demon" is necessary to understand living systems, and if this demon is some kind of information- or distinction-processing device, then at the heart of living systems there is a process of computation that is grounded in *abstraction*. For the naïve materialists of the past, this idea might have been shocking. However, the idea that a computational engine of abstractions lies at the heart of all living systems is a natural conclusion derived from Maxwell's demon.

Moreover, an intelligent creature observing the order pattern of equilibrium states may also conclude that when an increase in entropy is observed, without any clear external intervention, it is probably the result of a spontaneous process. Likewise, when a decrease in entropy is observed this is likely the signature of a non-spontaneous process where some work is done, energy is used, and entropy/heat is generated. A correlation is therefore formed at the most basic cognitive level by identifying the mutual information (i.e. correlation) between entropy states and related systems such as our mind. Let's explain this idea using a simple example.

An old person with dementia may exhibit a trajectory toward less distinct mental representations—for example, faces that they sorted into different "chambers" or named differently in the past are now undifferentiated. This would be like a situation where there was confusion over whether a molecule should be tagged as "fast" or "slow" in the context of Maxwell's demon.

Moreover, the mental decline of the person with dementia may be expressed by the fact that the variety of words that have been used to describe the richness of the world is now gone as the words have been collapsed into general semantic categories. For example, the variety of distinct names given to distinct objects may now be replaced by the words "This" or "That." When observing such a painful and gradual process of mental decline, we may conclude that it is a spontaneous process where the increase of entropy expresses the loss of distinctions. When coupling this process to the arrow of time, we may grasp the association between time, age, and mental decline. In contrast, a child learning to use language presents a trajectory from less to more differentiated representations. She learns to name different animals, such as a rabbit, kitten, or cow, and her developmental trajectory is indicative of work done, energy invested, and information produced. If our mind may be modeled as a kind of information-processing engine, then it is an engine of distinctions and is well aligned with the irreversible nature of the world through interactions and constraints.

Now back to the demon. The demon seems to threaten the validity of the second law, as it effortlessly turns disorder into order. However, as explained by Hemmo and Shenker (2012, p. 38), the laws of thermodynamic are generalizations robust "only in the appropriate circumstances for which they were meant to apply"; in the real world, statistical *fluctuations* are abundant, and standard thermodynamics has no conceptual tools to deal with these fluctuations. This may be a striking statement for those who portray a simple picture of the world; however, as explained by Susskind,[1] "Reality is messy and simplicity is imposed by our mind." This statement should be contrasted with that of Davies and Walker (2016), cited in the previous chapter: "the most powerful simplifications are abstract and generally hidden" (p. 2). In a messy reality governed by fluctuations, the simplicity "imposed by our mind" might be misleading even when we are trying to understand the most basic aspects of identifying distinctions and learning patterns through interactions, associations, and constraints. It is this messy reality that is the focus of my interest and not the ideal and imagined one, which is governed by the mathematical beauty of abstract simplicity. The messy reality can be approached through "abstract simplicity" produced by our mind as long as we use this simplicity as a general roadmap or analogy rather than as a law. A soccer team that operates according to a simple law is destined to lose, because in a competitive environment, its lawful behavior may be exposed and abused by its opponents.

The real world includes statistical fluctuations, which is a highly important aspect that adds complexity to our understanding of irreversibility. These fluctuations are a key source of information for our mind. In addition, the idea that the demon may work effortlessly is an ungrounded assumption. From any real-world computational perspective, the demon requires some energy in order to start doing its job. The first thing that it must do is to recognize a molecule as such. Otherwise, how can it sort out the fast-moving from the slow-moving molecules?! This process involves the ability to distinguish a signal from background noise.

It is argued (Collins et al., 2018) that the most powerful test statistic for differentiating a signal (*sig*) from background noise (*bg*) using some observable (Y) is the following likelihood ratio[2]:

$$L(Y) = p(Y|sig)/p(Y|bg)$$

This expresses the ratio between the probability of observing some "cue" (Y) given the signal and the probability of observing Y given background noise. When modeling the demon, we may hypothesize that any demon

---

[1] https://aeon.co/videos/the-whole-thing-is-a-monstrosity-how-a-symmetry-heretic-sees-the-universe.
[2] Known as the "Bayes factor."

seeking to identify a molecule should measure such a likelihood ratio. It therefore must invest some *energy* to sense the observable (Y), to form and maintain a *memory* device representing the probabilities, and to perform some *reasoning* through which the output $L(Y)$ is produced from the data. None of these operations can be done effortlessly. It therefore seems that from a cognitive computational perspective, the little demon cannot be used as an argument against the validity of the second law. However, the demon is an excellent context for understanding the operation of minds in an (1) irreversible and (2) fluctuating environment. If we learn *abstract* patterns from an irreversible and fluctuating environment, then these two aspects must be better understood. In this context, the fluctuations theorems (Sevick et al., 2008) may add depth to our thesis and progress our understanding of irreversibility and the mind.

Living with Fluctuations

Sevick et al. (2008) explain what it means to be irreversible:

> If the probability of observing all trajectories and their respective anti-trajectories are equal, the system is said to be reversible; on the other hand, if the probability of observing anti-trajectories is vanishingly small, we say that the system is irreversible (p. 605).

According to the second law, the probability of an anti-trajectory is zero, but it strictly applies to large systems or long timescales and not to very small systems. The fluctuation theorem of Evans and Searles (cited in Sevick et al., 2008) interestingly suggests that "as the system size gets larger or the observation time gets longer, *anti-trajectories* become rare and it becomes overwhelmingly likely that the system appears time irreversible, in accord with the Second Law" (p. 605). However, on a very, very small timescale and in a very small system, we may surprisingly encounter reversible processes. These scales are not directly relevant when modeling our mind. While some decision-making processes can theoretically be tracked down to the micro level, at which reversibility may be evident, for all living organisms, irreversibility is a matter of fact. The fluctuation theorem, however, may soften and deepen our understanding of irreversibility in the real world and may even apply to our cognitive processes, but what do we mean by anti-trajectories and time reversal? And how is the fluctuation theorem associated with these concepts? As I would like to argue, in its general conceptual form, the fluctuation theorem may have a very interesting interpretation.

In a text posted on ResearchGate,[3] the MIT physicist Andreas Mershin explains that "by definition, a reversible process is one that can be undone completely, leaving no trace of it having occurred." It is like the perfect crime scene, where the brilliant villain cleans the site and doesn't leave even a single trace of his atrocity. Such a process is perfectly symmetric and involves "the lack of any memory trace." Mershin further explains that upon reversal "it should be impossible to devise any experiment that could determine whether the process took place." This means, he further explains, "that every single bit of information that was flipped by such a process going forward, should be flipped back when the process finishes going in reverse." Postulating the existence of such a process, he further argues, is an unfalsifiable assumption because by performing such a process, paradoxically you would have no evidence of it at all. This means, he adds, that "asserting the existence of truly 'reversible' processes is no more than *an act of faith*" (emphasis mine).

Mershin explains the notion of reversibility in terms of memory and memory traces, which is not a trivial move. A reversible process is a process that can be reversed in time without leaving any memory trace. *Ipso facto*, an irreversible process is a process where memory traces are left. Paradoxically, if an irreversible process is a process where memory traces are left, then we can use these memory traces to go back in time. However, the idea of reversible computing, as pioneered by Fredkin and Toffoli (1982), proposes a different connection between memory and reversibility. To perform a reversible computation in which the input can be reconstructed from the output of a certain operation (e.g. AND), we can use an additional "logical gate" to achieve reversibility. Let me illustrate this idea through a simple example.

Let's assume that you use the operation of addition (i.e. +) and combine two natural numbers (i.e. inputs) to produce a third number (i.e. an output). Now, given your knowledge that two natural numbers have been added to form a third one, you are presented with the number 6 and asked to reconstruct the two numbers that have been used as the inputs. The answer could be 4 and 2, 3 and 3, and so on. From the computational perspective, *irreversibility is evident whenever we cannot restore the input from the output*. But what would happen if you added to your system another "gate," which is multiplication? When you add the value of two numbers, you also multiply them and keep the output so that the number of outputs is the same as the number of inputs. In this case, when you are given the output 6 and the multiplication output 8, then you know for sure that the added

---

[3] https://www.researchgate.net/post/Can_anyone_suggest_an_example_of_a_truly_reversible_process.

numbers were 2 and 4. The reversible process is performed not by erasing any memory trace but by using a new gate that in itself can be considered as a *memory trace*. Reversible computing is therefore possible but involves a heavy computational load on the memory of the system.

From my perspective, there are two important consequences that we may derive from Mershin's text. First, the idea of reversibility seems to be problematic and we may better talk about *degrees of irreversibility*, as actually proposed by the fluctuation theorem mentioned above. As stated in the fluctuation theorem, "as the system size gets larger or the observation time gets longer, anti-trajectories become rare and it becomes overwhelmingly likely that the system appears time irreversible" (Sevick et al., 2008, p. 605). Irreversibility is functionally associated with the system's degree of granularity (i.e. size) and observed timescale. We are *cognitively* able to track back the inputs that have led to an output, and this ability is somehow associated with the system's degree of granularity and timescale.

The second consequence concerns Mershin's statement that "asserting the existence of truly 'reversible' processes is no more than an act of faith." If you seriously think about the implications of this argument, then you may realize that this is a shocking idea. This is because *reasoning backward*, such as reasoning from effect to cause, is prevalent among human beings and to reason backward, we must assume some level of reversibility even if not a spontaneous one and even if not the one measured by the lack of any memory trace, as proposed by Mershin. In this context, an act of faith is not necessarily an ungrounded religious act such as a leap of faith. An act of faith may be interpreted as a process of probable reasoning. A bird anticipating the trajectory of a flying bug, in order to catch it, is involved in an act of faith as there is no mathematical logical guarantee whatsoever that the bug will behave in the anticipated manner. An act of faith seems to be the inevitable consequence of reasoning in an irreversible world characterized by statistical fluctuations. To explain why, let's take a step backward in order to gain a better understanding of what we are dealing with.

The Drill Sergeant and the Physicist

To recall, the second law of thermodynamics represents one of our most basic experiences as cogent beings: the observation that some processes are *irreversible*. And now for a simple example illustrating how deeply the notion of irreversibility is grounded in our most basic experiences. During my basic army training, my colleagues and I had to do some work under the guidance of a tough drill sergeant. In light of the need to complete a task within an impossible deadline, the sergeant had a philosophical insight, and he loudly shared it with us: "There is no motherfucker who can stop (the flow of) time." This drill sergeant was probably not familiar with the second law of thermodynamics and I doubt whether he had a high-school education at all, but by making this insightful statement, he represented the experience of time as we all know it. Regardless of our efforts, "time flies" and things change in an irreversible manner: the dead cannot be revived except in rare cases (e.g. Lazarus) and only through the miraculous intervention of God." History cannot be reversed and there is no point crying about spilled milk because the process cannot be reversed, and the milk cannot simply be squeezed out of the earth onto which it was spilled and returned back to its bucket. Although we can imagine a process through which the milk is squeezed out of the earth and reused, this cannot happen *spontaneously* but only by investing a lot of energy and work and not by replaying the exact chain of events backward. There is no point crying about spilled milk but there is a point in trying to understand a chain of events that will take place in the future. In our most basic experience of the physical realm, the arrow of time points to the order of things and represents the idea that irreversible processes cannot be *spontaneously* reversed. A spontaneous process is a process where constraints on a system's degrees of freedom do not exist and the system is freely allowed to explore its space. The fast-moving gas molecules in the demon's box can be freed from their cage by simply opening the door separating the two chambers. When this constraint is removed, the molecules are given the freedom to travel and visit each potential location in the box by finally maximizing their even distribution in a way that can be expressed as maximizing the entropy of the system. In this realm, no difference can be found between past, present, and future. A change toward an increase in entropy may therefore indicate that some constraints have been removed and a decrease in entropy may indicate that some constraints have been imposed. The exact nature of these constraints may be revealed, however, when we examine the interaction between systems and identified correlations, as explained before. We all swim in a river heading in a specific direction, but interestingly in this river there are fluctuations that may be used by intelligent systems that may function as little demons and by a small number of demons working together to achieve the same goal. Each demon invests some energy in identifying differences, and this energy is used to identify other differences and similarities that, in turn, may be used to maintain the demon's own cognitive system.

Previously, I defined an irreversible process in thermodynamic and computational terms. However, to model irreversibility, we may try to adopt a more general and abstract notion of irreversibility. A process may be defined as time *reversible* if for every N the series $\{X(t_1), \ldots, X(t_N)\}$ and $\{X(t_N), \ldots, X(t_1)\}$ have the same *probability distribution* (Weiss, 1975). Please notice that here our main emphasis is on a probability distribution rather than on available energy or entropy. Therefore, the *irreversibility* of a dynamical system is evident when, under an imagined time-reversed transformation (a procedure that in itself must be well defined), the reversed image of the original time-series distributed differently or said differently, the distributions are asymmetric.

When talking about irreversibility, it may sometimes be easier to think in terms of symmetry and symmetry-breaking. This idea is applied to the context of traveling back in time and space. We may imagine an irreversible process as one in which we travel through a wood but cannot return back along the same path by tracing our own footsteps. In *The Garden of Forking Paths* (Borges, 2018), there is no way to track back our own footsteps. Now, the irreversibility of a process means that the transition probabilities are not time reversed when comparing one state to another. The probability of being an old man at t_{80} given being a baby at t_8 is very high while the reverse transition probability of being an infant given first being an old man is null. Exceptions, such as *The Curious Case of Benjamin Button* (2008), emerge from the artistic imagination rather than from the way the real world usually works.

As $p(X(t_N)/X(t_{N-\tau})) \neq p(X(t_{N-\tau})/X(t_N))$, our ability to reason backward may be severely limited. In such a world where reversibility doesn't actually exist, we may ask why we care about it at all and how various organisms—from bacteria to human beings—compensate for this lack. Here is the idea that I would like to propose. Organisms require reversibility in the sense of *backward reasoning*, which is detrimental to their ability to learn, survive, and adapt. Memory, in the sense of encoding information to produce general representations, is the solution to the need to perform backward reasoning in an irreversible world in order to impose some order. This idea is further developed in the next section and its relevance to understanding small systems will be described step by step.

Backward Reasoning as an "Act of Faith"

Let us return to Mershin and his argument that "the existence of truly 'reversible' processes is no more than an act of faith." This is a shocking idea as "reasoning backward," such as reasoning from effect to cause, is very common among humans. For example, in his book *The Sherlock Effect*

(2018), the forensic pathologist Dr. Thomas Young criticizes the way "doctors and investigators disastrously reason" (the subtitle of the book) by reasoning backward. His argument and criticism are extremely simple. Reasoning backward involves "describing the multiple steps in the past that led to the result" and "there is a simple explanation why backward reasoning does not work. For any result, any set of clues, there may be numerous possible 'trains of events' that could explain the result" (Young, 2008, p. 2). This idea appears in numerous places, such as in the novel *The Guest Cat* (Hiraide, 2014), where the author explains that the way events unfold depends on an infinite number of mundane factors. Young is not a physicist, and his work has nothing to do with thermodynamics or irreversibility. However, his statement is deeply grounded in the laws of nature and in a possible cognitive computational approach to reversibility. His solution to the Sherlock fallacy is to substitute backward reasoning with the careful analysis of clues and their coherence, or the way they form a pattern that may be used to understand what really happened. It seems that in order to find some order in an irreversible world, we should follow this pathologist's advice. Memory, to be discussed at length in the following sections, is a device that keeps track of various "information clues," and patterns are formed by connecting these dots together.

As time unfolds or as the size of a system increases, the probability of observing a reversible process exponentially diminishes, and so does our ability to reason backward and to track the *exact* path that led to a specific output. To learn abstract patterns, it seems that we must trace our footsteps or the path leading to a specific output. As in an irreversible world such a process is extremely limited, we are facing a problem. As we can see, irreversibility, a concept from physics, may be interpreted as having a deep *cognitive computational* meaning. Our understanding of this process may be enriched again by the idea of constraints.

A reversible process is a process where the forward and the reverse paths must be the same. When we seek the exact path that led us to a present output, we actually seek the chain of events (i.e. the order of things) that occurred, given some constraints (e.g. the constraint of causality). This is a classical context of combinatorial optimization and in this context a problem, such as finding the reverse path, is considered to be "hard" if the solution's computing time grows exponentially or faster as a function of the number of components. It is possible that performing a reverse search for the exact path that led us to a certain time point would be such a problem, and that this is the reason why some kind of cognitive fluctuation theorem may nicely describe the difficulties inherent in any attempt to reason backward.

This logic of irreversibility is evident in the context of finding a missing person too (Holmes, 2016). In such cases, the first 72 h are considered crucial for finding the missing person.[4] The reason this period of time (i.e. 72 h) is deemed critical is mainly the potential for loss of evidence, whether as a result of loss of human memory (i.e. eyewitness accounts) or the loss of forensic information, such as a blood spot being washed away by rain. As explained by a former FBI agent: "As time goes on, there are fewer '*breadcrumbs*' to follow" (emphasis mine). *Reversibility is therefore deeply associated with the existence of memory traces*, or what the FBI agent describes as "breadcrumbs."

From a cognitive computational perspective, which guides my understanding of small social systems, reversibility concerns our ability to identify the input(s) that resulted in a specific output, using some breadcrumbs. But why are there fewer breadcrumbs to follow? The answer must somehow involve the ideas of *information* and *memory*. These two terms are not canonical parts of physics, but when we try to explain foundational issues concerning reversibility, they pop up again and again. The next section delves deeper into the concepts of reversibility, memory, and information.

Memory, Reversibility, and Information

We usually think about memory in anthropocentric and mentalistic[5] terms. However, memory doesn't have to be considered as purely mentalistic (or psychological) concept. Memory is just a registration of past events on some medium and remembering is just the use of these traces. Memory doesn't keep a copy of the past or simply revive it. As we have learned, the dead past cannot be revived. Memory is therefore just a way we record, store, and recover/use *traces* of the irreversible past that has been lost forever.

Here is a highly important point for the thesis that I would like to propose: *the basic experience of irreversibility has a deep cognitive sense that seems unavoidable. We cannot revive the past and therefore our ability to learn from it leans on our ability to walk backward by using some memory traces, as expressed in the formation of abstract representations.* More accurately, the memory traces are *signs* that allow us to learn order patterns by avoiding a real and impossible reversible process. The memory traces are "signs" as they are

---

[4] https://abcnews.go.com/US/72-hours-missing-persons-investigation-critical-criminology-experts/story?id=58292638.

[5] The dichotomy of material and non-material is old-fashioned and meaningless in the current science. I use this statement only to address some false conceptions still existing in the social sciences and humanities.

not the past itself but just a device that points to a past event through some kind of association. *Semiotics*—the study of signs and signification (Danesi, 2004; Sebeok & Danesi, 2012)—is therefore highly relevant to studying our mind.

Memory, as a register of the past, is deeply connected with the future. In the context of cognitive psychology (Conway et al., 2016), it is proposed that "remembering the recent past and imagining the near future take place" in a "remembering–imagining system" (p. 256). To "imagine" or predict the future, we must remember the past. The irreversible nature of the world imposes constraints on the order of things. However, to learn these order patterns as if we were "demons" observing them, we must perform some analog process to reversible computing by restoring the "inputs" that preceded the output. In this context, remembering the past isn't reviving the past; rather, it is the ability to use memory traces to learn abstract patterns and order patterns. A mind operating in an irreversible environment would find out that history can teach us nothing, as history doesn't repeat itself, and that there is no point whatsoever learning from experience. A mind operating in a fully reversible environment is hard to imagine. The best engineering solution for a mind living in an irreversible and fluctuating world and seeking to learn from experience would be to use a memory system that compensates for the lack of reversibility.[6] This memory is actually the registration of traces that may function as signs to re-present the dead past in a way that is usable for the formation of tailor-made solutions in real time. Here you may start grasping the way in which the cognitive and social are connected. For a living being to maintain memory, a system of signs is a must. As signs are inherently social (Volosinov, 1986), our most basic cognitive processes are social through and through. This is another conclusion that I find striking: not only that at the heart of living systems there is a process involving computation and abstraction, but also that this process is social through and through!

Memory is not an exclusively mentalistic concept and, surprisingly, matter expresses different forms of memory too (Keim et al., 2019). Pressing your thumb against a piece of clay leaves a mark, a callus on your hand is a memory trace of hard work, scars are memory traces of traumas, and a wake in the water indicates that a ship has passed. These are simple memory traces disconnected from any cognitive agent. A mark on a piece of clay is not a memory trace unless someone uses it to recall a past event.

---

[6] I thank my colleague the physicist Boaz Tamir for helping me to sharpen this idea.

Entering the cognitive dimension of memory doesn't necessarily take us too far from the physical realm. For example, a friend of mine, the anthropologist Zvi Bekerman, told me many years ago about a meeting he had with the biologist Humberto Maturana, who explained to him the "material" nature of memory through a concrete example. This concrete example takes us outside the simple physical realm, but not far enough to lose sight of it. The example is as follows. The wheel of a car hits the pavement and, as a result, is twisted. When the car drives above a certain speed, the wheel starts shaking. The wheel is contextually "remembering" its accident and this remembering can allegedly be described with no reference whatsoever to human mentalistic concepts. Speed is the context in which the wheel's memory of the trauma is enacted. The twisted form is a registration or trace of the traumatic encounter with the pavement and the shaking expresses the enactment of the trauma.

We may therefore conclude that when the behavior of systems is coupled (e.g. the encounter between the wheel and the pavement), it may leave a trace at least on one of the systems (e.g. the twisted form of the wheel). Context (e.g. the high speed) provides the trigger through which the behavior of one system is regularly constrained (e.g. whenever we drive above a certain speed, the wheel start shaking) in a way that is indicative of the past interaction with the other system. Whenever an interaction leaves a trace that is expressed within a triggering context, we can talk about memory.

It is important to emphasize again that the trace *is not* the past itself. The twisted wheel *is not* the violent encounter between the car and the pavement. It is a trace or a registration of this encounter, which in itself cannot even exclusively function as a memory trace. It is a registered mark of the incident that can function as a memory trace, or as a sign, only under certain circumstances (e.g. high speed) when it produces a regular behavior that points back to the traumatic past.

In other words, the trace can function in a memory system only when it starts functioning as a *sign*. Why is it so important to insist on this point? Because we have just shown that memory involves some kind of a *re-presentation*. The twisted form of the wheel is not a literal image of the trauma. It is not a film taken and showing the violent encounter between the wheel and the pavement. The twisted form is a re-presentation of the trauma, just as a scar is a re-presentation of a physical trauma. It is abstract in the sense that it lacks the concrete details of the past event to which it points. From examining the trace, we don't know who drove the car, where or when exactly the encounter took place, and so on. The trace is just a simple representation, a mark, a distinction, which can only be used under certain circumstances; the car "recalls" or behaviorally expresses its trauma as the constrained behavior (e.g. the wheel isn't free to "choose" whether to shake or not if it is above a certain speed) evident in a given context. This is of course a much more complex description of remembering that is far beyond the physical examples of memory described above. In contrast with a mark left on a piece of clay, the cognitive systems discussed in this book are able to use a memory trace to *monitor* and *regulate* their behavior.

To the best of our knowledge, a memory trace left on a piece of clay isn't proactively used by the piece of clay. Organisms, in contrast, use the complex memorization process described above to regulate their behavior. A memory trace may be registered even in a bacterium's "mind" (Lan & Tu, 2016) up to the genetic level and in such a way that it changes the bacterium's behavior. In fact, there are some engineering solutions which propose that memories can be stored and written in the DNA of living cells (Farzadfard et al., 2019; Tavella et al., 2019). However, the registration of memory on the individual level of analysis should take into account that memorization is a collective activity and that there is no private memory, just as there is no private language. Memory is always a *collective memory* because when we form our representations, we collectively form an abstract signification system in order to overcome the limits of our cognitive processing and our limited individual perspective. What holds for collective remembering among human beings (Middleton & Edwards, 1990) also holds for bacteria (e.g. Ben-Jacob et al., 2004). We may therefore conclude by understanding that memorization is a process necessary for learning patterns in an irreversible world, specifically through some form of collective activity that generates abstract representations through a social system of signification. This is a highly important point that I would like you to keep in mind.

Memory traces are physically grounded and therefore they naturally decay with time, such as when old symbols written on a piece of sandstone are lost as a result of wind, dust, and friction. When we try to reason backward, we actually attempt to reconstruct the path that has led to a current state (i.e. the output). As time unfolds and/or the size of a system increases, the trails of breadcrumbs through which we have to search backward become bigger and bigger (like spreading tree roots) and more intensively exposed to natural decay. Each junction in the roots is a state that is the computational output of previous states, and the complexity of the roots grows exponentially. When reasoning backward through memory, we therefore face the problems of (1) natural decay and (2) the exponentially increasing number of potential paths. Irreversibility may therefore be expressed in *cognitive* terms as the difficulty we experience reconstructing a given path under both real-world and cognitive computational constraints. While irreversibility is primarily a physical phenomenon, it has a clear cognitive correspondence in the mind of all living creatures as systems grounded in the physical realm. To better understand why, we need to further elaborate the computational perspective on irreversibility.

Bennett and Landauer (1985) discuss irreversibility through the physics of computation. First, they argue that they use the term *information* in the technical sense of information theory. They also suggest that "information is destroyed whenever two previously distinct situations become indistinguishable" (p. 48). *Ipso facto*, we can understand that information is equivalent to a difference or distinction, as previously proposed by Rashevsky and others (e.g. Bateson, 1972/2000). The basic unit of information, at least conceptually, is a distinction, and when two distinct states become indistinguishable, information is lost. The consequences of this idea are very interesting. For Shannon, information is a measure of surprise, which is expressed through the distribution of a random variable. However, if the basic conceptual unit of information is a distinction, then information appears when distinctions are added or lost at a system's different levels.

Distinctions may be added or lost on different levels of analysis, such as when the distinction between Kitty the cat and Jackey the cat may be lost in favor of the more general concept of a cat (Tamir & Neuman, 2016). Losing on one level for a gain on another level may be a beneficial cognitive strategy for several particles collectively operating. This idea is not fully aligned with Shannon's idea, which is a point that deserves some consideration. It may become clearer when considering a collective representation formed when two particles produce an output from their two individual representations. When forming a collective representation, some information or distinctions may be lost as an inevitable part of the computation process, but the more general and abstract output may have the benefit of supporting the communication and cooperation of the particles. Trying to communicate while holding in mind our own unique idiosyncratic representations of a cat would be an impossible mission, just as it would be to try to communicate through some kind of private idiosyncratic language.

Bennet and Landauer further explain that the *erasure* of information, or the loss of a distinction, is accompanied by the dissipation of energy (i.e. entropy or heat). A simple example that we have previously used may illustrate this point. A tennis ball is dropped from either 1 or 2 m from the ground. In a situation where there is no friction and the ball is perfectly elastic, an observer will always know whether the ball was dropped from 1 m or 2 m, as the height to which the ball bounces will be indicative of its original state. In a frictionless world, information may be kept forever, but that information has no value as it exists in a void. Information is informative only for an observer, whose existence cannot even be imagined in such a reversible world, as illustrated through the demon thought experiment. Minds are evident only in situations where irreversibility and loss can be observed and used.

In contrast with a frictionless situation, when friction exists, the ball would release some energy whenever it hit the ground until it finally came to rest, indicating nothing about the height from which it had been dropped. The same is true, Bennet and Landauer argue, for a process of computation, which is an operation that produces an output from inputs. As explained above, even if we observe the output 6 and know that the binary operation is +, we still cannot reconstruct the exact values of the inputs. We can produce 6 from 3 + 3, 2 + 4, and so on. A process is an irreversible process whenever information or distinction is destroyed during a process of computation and this process exists in an observer's mind, as illustrated by Bennett and Landauer's ball example. Irreversibility is therefore understood as a process of computation where the input cannot be reconstructed from the output, where some distinctions are lost, where information is lost, and where work is done, and heat is released as we erase some information. Therefore, irreversibility is deeply associated with loss, as we have learned from *The Hobbit*, but Landauer and Bennett ground this loss in physical and computational processes. We may now better understand why our mind is so deeply grounded in irreversibility. As the basic unit of the mind is "a difference that makes a difference" (Bateson, 1972/2000), and as irreversibility means the loss of distinctions during a process of computation, information is formed as an abstraction of lost distinctions.

Whether information is lost as a result of friction or erasure is of no importance to our thesis as the most general lesson is that irreversibility is framed in terms of information loss. *Our ability to learn, adapt, and survive is deeply connected with loss*. To repeat our thesis, loss is by definition inevitable in an irreversible world represented by a mind performing irreversible computation. An organism striving to learn from the past—even if to distinguish signal from noise—must therefore compensate for this loss through memory, which is an engineering solution that involves the formation of abstract representations in order to reason backward and learn from experience in an irreversible environment. However, in an irreversible world, even memory is prone to loss or decay and cannot serve as a full record of the past. Loss is inevitable but manageable to a certain degree.

也 池

Fig. 3.2 Chinese characters. *Source* Wikipedia

What is the solution to the tension between the need to memorize and form abstract "mental" representations and the inevitable loss? An answer may surprisingly come from Seneca's *De Brevitate Vitae* (*On the Shortness of Life*) (2004). In this book, the great thinker (who lived in the first century CE) explains that there is nothing that time doesn't destroy or erase. However, he adds, *ideas* perpetuated by wisdom are immortalized and cannot be damaged by time. This is a very interesting and deep thought. While information may be destroyed on the micro level, ideas may better survive as they exist on the macro level. But what does Seneca actually mean by "idea" and how is this idea different from information?

We have considered information in basic terms as distinctions, but ideas are not simple distinctions and living systems are not simply information-processing systems but "meaning-making" systems (Neuman, 2004). Have a look at these Chinese characters (Fig. 3.2):

You may easily notice that they are different. However, if you are not familiar with Chinese characters, then you will not understand their meaning. Distinctions are meaningful only when correlated with something (e.g. references) that loads them with value (Neuman, 2017) and makes them useful for action.

Ideas are basically formed as abstractions of differences and as abstract patterns of differences. When a difference in available energy is sensed by a colony of bacteria, they experience the idea of "food." It is the general and abstract pattern of all possible gradients that they have experienced in the past and that is encoded in their genes as a collective. *Ideas are therefore abstractions formed and maintained through collective activity over time.* There is no idea of God without a collective of people maintaining it, and there is no idea of a face without a collective of neurons bringing it into life.

The idea that I would like to propose is that small social systems, such as families, soccer teams, and jazz trios, are exceptionally good at forming ideas and flexibility using them in practice, in real time, while optimizing adaptability. This highly important thesis may raise a brow; what is the "idea" used by a soccer team, a family, or a jazz trio? Can we think about these systems as some kind of collective form of a philosopher? The answer is clearly negative. The major goal of a soccer team is to score as many goals as it can while keeping itself safe from the mirror goals of the competing team. The aim of a family, or at least one of its major aims, is to care for its children. The aim of a jazz trio is to create an aesthetic experience for its audience. To address each of these challenges, the respective system must form an abstract and shared representation through which it regulates its behavior. This is not a conscious, explicit, and well-formalized idea. When a soccer team tries to coordinate the behavior of its players in order to maximize communication and performance, the idea is the abstract, dynamic, and adjustable pattern evident in the performance of the players. See, for example, Marcelino et al. (2020), where it is found that the coordination tactics of a team are not random but expose a pattern. This pattern is a kind of idea, and the way we may think about it will be clarified when I introduce Rashevsky's notion of *topological information*. For now, you can just ignore the dogmatic and philosophical ideas you may have learned concerning what an idea is all about. Surprising as it may seem, in nature ideas exist as information "in between" interacting particles. At this point we may ask why small social systems may be so effective in flexibly forming and using such abstractions. While a simple answer is that the number of constraints forced on the collective of a small number of particles is optimal for their achieving some goals, I would like to look further and investigate an additional and interesting hypothesis.

Irreversibility, Leaves, and Decay: Learning from Loss

Let us discuss the notion of decay and the increase in entropy that we observe almost everywhere. A memory trace is prone to decay, but the process of decay has its own logic and by forming the idea of this process we actually learn about the constraints that operate on the system. My thesis is that a group of people may generate abstract representations as a by-product of the "information decay" process accompanying the shift from the individual to the collective level of organization.

To illustrate the idea of learning from decay (increase in entropy), let's think about movies and graphic images of *corpse decomposition*. If you have ever watched a horror movie, then you have probably been scared by images of decaying corpses—specifically, those that return to hunt the living out of some vengeful or other morbid motivation. These graphic images are not limited to horror movies; they have appeared to powerful effect in the TV series *Game of Thrones*, where the army of the dead is an illustrative example. Now, think about the decomposition of a corpse. When the heart stops pumping blood, oxygen doesn't reach the cells, and organs that are more

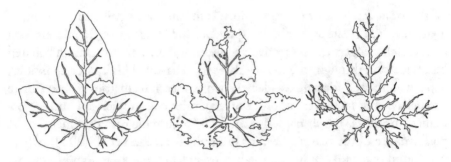

Fig. 3.3 A decaying leaf. *Source* Author

heavily dependent on the supply of oxygen, such as the brain, are the first to decompose. The immune system, another complex and energy-consuming system, stops functioning too and the result is the release of hungry bacteria, which start annihilating the intestines and the stomach. The more complex organs and systems are the first to go, and therefore the increase in the entropy of the human corpse exposes the *constraints* operating on the body. Given this process of decay, the last aspects to "survive" expose the more general form of a human body. This point may be more easily explained through the decay of leaves. Figure 3.3 is showing such a process of decay.

Following the process of decay from left to right, you can easily see that the increase in entropy isn't evenly distributed across the leaf's overall structure. What we see is that the margin or the *boundary* of the leaf is the first to structurally change. *Ipso facto*, we may produce the notion of the leaf's abstract form through this process of decay. Here is the explanation. Observing the leaf, it looks as if someone took a bite out of it. The change in the boundary is more evident in the middle image, where it has begun to look like the coastline of England. The leaf's boundary now seems to have a kind of fractal structure.[7]

The boundary is one of the essential properties of any pattern, and decay—an irreversible increase in entropy—is always accompanied by the decomposition of a boundary. When we observe the decay of the leaf, we actually learn the abstract pattern of a leaf through this decay and the exposure of the boundary. A very simple heuristic that may be applied to such a

---

[7] The idea of a fractal dimension is beautifully explained at https://www.youtube.com/watch?v=gB9 n2gHsHN4. By watching this explanation, you may better understand why the increase in entropy is (nonlinearly) accompanied by an increase in the roughness—or fractal dimension—of the leaf's boundary. The more constraints are released from the system, the rougher it may become.

case is to observe the pattern of decay and picture the lowest fractal dimension through which the boundary may be represented so that the observed object can still be sufficiently distinguished from other objects (e.g. "decaying" basketballs).

The famous architect Christopher Alexander describes the properties of a pattern in his insightful book *The Nature of Order* (Alexander, 2002). One of the properties is the *boundary* and, as we can see above, the boundary immediately changes during decay. As we see from the rightmost picture, the leaf's "skeleton," its midrib and veins, is the last to go. The exposure of the leaf's skeleton also exposes the leaf's symmetry. As argued by Alexander, local symmetries are another property of a pattern and the leaf's symmetry is kept even over progressive phases of the decay. *Irreversibility, as we can see, may help us to expose the general pattern of some observable.* We have all seen the collapse of a structure, such as a panicking audience, and such increases in entropy may teach us about constraints, their release, and the trajectory of a system under the release of constraints.

Interestingly, this process of decay is captured by the idea of *information compression* too. Compressing an image through *lossy compression*, where some information can be disregarded, results in redundant information being lost just to keep the "essence." This essence captures the "idea" of the thing that is compressed. Have a look at these two images (Fig. 3.4):

On the left is a black-and-white image of a maple leaf and on the right is its low-resolution representation. By observing the irreversible process of decomposition or decay, we may learn not only about the order of things but also about the idea, form, or essence behind them. If you think about it, then you may realize how death is associated with the emergence of ideas. A mind coupled with the physical environment may learn a lot from irreversible processes and extract general patterns through this observation. When we observe an irreversible process, such as a process of decay, we learn about the

Fig. 3.4 A compressed representation of a leaf. *Source*: Author

constraints operating on the observed system and are potentially capable of extracting its "essence," which is its abstract representation as formed by the collective.

When a collective of individuals computes, the output inevitably involves the loss of information. However, if the particles communicate their outputs to form a collective representation, then this collective representation is necessarily some kind of abstraction. Loss of information is inevitable, but, when communicated and coordinated, it may lead to an abstract form of collective representation.

Back to the leaf and the logic of loss. I would like to add some depth to our understanding. What is lost is not only information in the form of *distinctions*. What is also lost is some kind of information about the *structural and functional integrity* of the system that we are observing (e.g. the leaf). To elaborate this point, I would like to turn to an insightful paper written back in the 1950s by Nicolas Rashevsky (1955). Understanding Rashevsky may help us to comprehend the notion of ideas as information "in between" and the way small social systems generate ideas and use them in real time to create tailor-made solutions.

Information "In Between"

Rashevsky was interested in the way organisms produce a high level of complexity from a limited number of basic building blocks. Considering complexity in terms of distinguishable units, there seems to be a quandary in the way complexity is evident in a system with a limited number of building blocks. Rashevsky's (1955) insight was that: "It is, however, possible for an aggregate of identical, physically indistinguishable units to have a large information content. The units in this case, though indistinguishable physically, are different through the difference of their relations to each other" (p. 231). Information, as Rashevsky realized, may exist in the *relations* between the elements rather than in the diversity of the elements per se. He illustrates this idea through an analysis of the information content of graphs (Fig. 3.5):

Let's look at graph a, for example. We have two units. Units 1 and 2 are indistinguishable and therefore the information content (i.e. information entropy) of the graph is 0. In contrast, graph b presents two indistinguishable units that are *topologically* different. Unit 1 is the end point of the directed relation whereas unit 2 is its beginning. The information content of this graph is 1 bit, as we have to ask only one binary question in order to know which is unit 1 and which is unit 2. Now we move to another graph, which has three

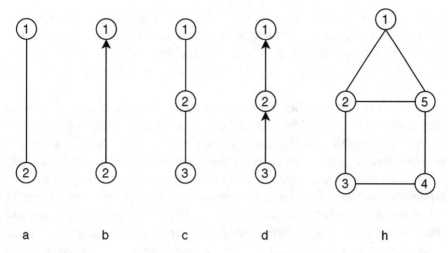

Fig. 3.5 Topological information. *Source* Author

rather than two particles. Graph c shows us three seemingly indistinguishable units: 1, 2, and 3. Unit 2, however, is different from the others because it has a degree of 2, meaning that it is connected to two other units. Units 1 and 3 are topologically undistinguished as each is connected to only one unit. The probability of being a unit of degree 2 is therefore 1 out of 3 (i.e. 1/3) and the probability of being a unit of degree 1 is therefore 2 out of 3 (i.e. 2/3). Shannon's information entropy is therefore calculated as follows:

$$H = 0.66 * \log \frac{1}{0.66} + 0.33 * \log \frac{1}{0.33} = 0.92.$$

If we make the three points fully distinguishable, as in graph d, then the information content of the graph increases to become 1.59 (as we have three equally likely options).

Now have a look at graph h. Units 1, 3, and 4 each have degree 2 while units 2 and 5 each have degree 3. However, unit 1 can be distinguished from 3 and 4. The reason is that unit 1 is adjacent to two units of degree 3 while units 3 and 4 each have neighbors consisting of one unit of degree 3 and one unit of degree 2. The entropy or the information content of this graph is therefore 1.53. As we can see, the information of the graphs is determined by their topology, meaning the relations hold between the units/particles. Rashevsky described this as the *topological information content*, but the overall information content of a graph may be determined by both the distinguishability of its units and the distinguishability of their relations. At this point, we may understand that symmetric structures have a higher level of entropy, meaning that the individuality of the particles or their locations is less easily discerned.

Rashevsky's idea may help us to understand several observations. First, it may explain Galton's (1889) observation that "the huger the mob, and the greater the apparent anarchy, the more perfect is its sway" (p. 66). How is it possible that the greater the anarchy, the more "perfect is its sway"? Isn't that a contradiction? Let's imagine a group of football fans after they have experienced a bitter loss. Drowning their sorrows in alcohol, they start chanting together and by chanting, simple symmetric relations are established between them and the individuals form a symmetric structure of interacting human beings. At this point, the entropy of the mob's structure is maximal, and the fans have lost any signs of individuality. The collective of individuals has moved into a new phase where it functions as a novel entity with a unique logic and dynamic. When individuals are part of a mob—indistinguishable particles in a symmetric structure—their mutual constraints make the whole simple to understand. It is like the *phalanx*—the rectangular mass military structure formed by the Roman army (Fig. 3.6):

Individuality has no place in a phalanx. It is a structure that imposes strong constraints on the behavior of the individual soldiers. It has, of course, the huge benefit of organized and ordered behavior. There is no place for anarchy,

Fig. 3.6 The phalanx. *Source* Wikipedia

and this brings the benefits and costs of an ordered structure. But the mob formed by the football fans is not quite the same, as it is formed from the *bottom up* rather than as the result of the top-down exercise of commands. However, they are similar in terms of their symmetrizing relations. If enough people adopt symmetrizing relations in a bottom-up manner, a mob will pop up in front of your eyes, whether a mob of football fans or a mob of academics adopting herd behavior.

Rashevsky's idea of topological information content may also help us to understand the potential novelty of interactions. When a small number of particles interact, they may form novel behavior. We can think about this novelty in terms of a new graph formed in between the particles. Take a look at Fig. 3.7.

The leftmost graph is merged with the middle graph to produce the rightmost graph. In the leftmost graph, node A is not distinguished from the two other nodes; however, when the leftmost graph is merged with the middle graph, a new graph is formed where A is distinguished from the rest. As argued by Cohen (2016) and following the work of Atlan and Cohen (1998), "complexity increases because the more information there is, the more likely that that information will become available for new engagements and, hence, for increased complexity." What we see in the above figure is that the complexity of the shared representation becomes higher from left to right and the novelty of the new structure is even higher.

The same logic of topological information may be relevant to understanding small systems. Small systems composed of a limited number of distinguished particles may change their information content only by changing their relations. Although a coach may substitute new players into a game or remove from the field those who fail to contribute, the major source of information content and complexity in a soccer game is the relations between the players and the way these relations change over the course

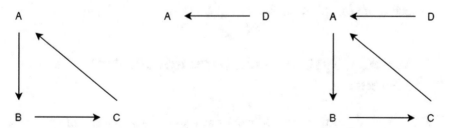

Fig. 3.7 New information from merging graphs

of the game. This idea will be illustrated in the second part of this book. For now, I would like to return to the decay of the leaf and to how ideas may be generated by observing irreversible processes.

As I have suggested, memory in the form of an abstract representation is formed when some information is lost but the loss isn't evenly distributed across the parts of the system, scales, and time. That is, a lot of information may be gained if we examine the pattern of loss. This is the way meaning, or what Gregory Bateson (1972/2000) described as a "difference that makes a difference," is formed. Learning how a leaf decays allows us to learn the general idea of a leaf. Learning about the decomposition of a human corpse allows us to learn that the more complex organs and systems are the first to go. When we understand this process, we may understand why bones have long been considered the infrastructure from which the dead would revive. In the Book of Ezekiel, we read this prophet's vision of the *Valley of Dry Bones*. In the vision, the prophet stands in a valley full of dry human bones. The bones connect to form human figures, then become covered with tendons, tissues, flesh, and skin. The resurrection of the dead starts with the bones rather than with the heart (the first organ to be formed in an embryo), the kidneys, or the brain. As the bones are the last thing to decompose during the decay of the human body, they are considered to be the solid ground from which the dead can be revived. The bones are the last part to lose their structural integrity, and by examining ancient paintings (such as those found in Magura Cave, Bulgaria, or the Bhimbetka rock shelters, India), you can see representations of human beings as bony creatures or silhouettes whose essence is captured by their boundary lines, largely determined by the structure of their skeletons. The general and simplest idea of a human being is also grasped by children, whose grotesque drawings of the human figure are simple and preserve general boundaries and relations. The information content is high, but the complexity and redundancy are low. To conclude, ideas are formed by relying on the natural order of the world, but when we dig deeper things are much more complex, as always…

Small Social Systems Generate Information in Between

The first hypothesis that I would like to discuss is that small systems are good at generating ideas because a lot of information exists in between the individuals of which they consist. Some of this information is lost when processed in between the particles, and the information formed in between the particles is what ideas are made of. Previously, I explained how information exists in the relations formed in an observed system. Relational information exists on the global level but also at different levels of analysis and time. However, information also exists in the topology of a small system, where representations communicated between the particles may change the *relations* and the *identities* of the composing units. By collaborating, the particles form a kind of dynamic graph where information not only exists at the particles' level of analysis (i.e. in their minds) but also in between them, in the topology of the graph, which is recursively updated and represented in the particles' minds. This "topology," or relational configuration, is formed by the particles as they communicate their various non-coinciding "perspectives." This process of perspective-taking should not mistakenly be understood as exclusively human. The perspectives that I have attributed to each particle are actually their individuality, and individuality is taken into account even in the study of *Escherichia coli*, where the word "individuality" is used to describe "genetic and nongenetic variation" that must be taken into account when a collective behavior is formed (Fu et al., 2018, p. 1). The "individuality" each particle possesses should not be considered in narrow anthropocentric terms. It exists even for *E. coli*. The individuality is actually expressed in a unique *perspective*, meaning that the representation produced by each particle differs from the perspectives of others as a result of genetic and nongenetic factors, such as the specific position of each particle and the unique point of view that position provides.

The idea that different perspectives are communicated and integrated to form a graph where information exists in between the particles should not be considered in anthropomorphic terms per se. Take the immune system as an example. The immune system, as a complex cognitive system (Cohen, 2000), cannot use only ready-made templates based on each and every opponent that it has met in the past (and therefore might meet again in the future). Keeping track of each opponent might overload the system's memory and therefore this might be a limited strategy. Moreover, building a repository of your previous opponents is not enough as along the way you may meet new opponents you have never seen before or ones that have mutated to avoid being identified. To address the challenge of immune recognition without falling prey to the fallacy of using a general recognition template, a different approach is required. Immune recognition, as explained by Cohen, is an emerging phenomenon resulting from what Cohen describes as the "co-respondence" of the particles composing the immune system (B cells, natural killers, etc.). The idea of an "antigen" triggering the response of the immune system is therefore formed in between the particles similarly to Rashevsky's topological information.

The idea formed by a system is actually the dynamic topological information that is expressed through the communication patterns of the particles. But this abstract pattern, which emerges through the graph topology, cannot explain the flexibility of the formed idea or representation. In the context of tailor-made solutions, the meaning of flexibility is that by observing a situation from its narrow and limited perspective, and by trying to communicate this perspective and integrate it with the perspectives of other particles through feedback loops, the system is pushed to a "solution" that maximizes its freedom of prejudice given a limited number of constraints enforced by the different perspectives. This is Edward Jaynes' idea of *maximum entropy*—discussed in the next section. The idea is brilliantly simple, and it can be explained through a concrete example.

On Diets, Nazis, and Entropy

Let's assume that as a member of a wealthy society, you are unsure of the most appropriate diet to follow. You are baffled by the media and various "experts," and cannot decide what to eat and how much. Should you be a fruitarian? Adopt a ketogenic diet and eat plenty of steak and fat? Should you consume coffee and how much? These are all bothersome questions for the perplexed of wealthy societies. We can think about this problem in terms of finding the best *distribution* of food, which relates to the probabilities of each potential food in our diet, from grains to meat. Which distribution should you chose? One that includes 30% whole grains or 15% whole grains? One that includes zero fat or 30% fat?

Edward Jaynes was a physicist who made an interesting proposal that is in line with *Laplace's principle of indifference*. Jaynes' (1957) proposal was that among all possible distributions available for your choice, you should choose the one that maximizes the entropy of the distribution under some few chosen constraints expressing your limited knowledge of the situation. The constraints are simple limits that we impose on the solution space when trying to identify the best distribution.

For example, if you observe a sample of people with a certain average height and would like to guess the distribution of height in the population from which this sample was taken, then according to Jaynes you should first choose the distribution of height that matches the average you have observed in the sample. For example, the average height in your sample may be 170 cm and there are many possible distributions that may be characterized by this average. A homogenous distribution of heights may suggest that all of the people in your population, regardless of age or gender, have exactly the same height, which is 170 cm. In such an imagined population, babies are born with a fixed height of 170 cm and remain that height until their death. Jaynes proposed that among all possible distributions limited by our constraints/knowledge, you should choose the one that maximizes the entropy or the heterogeneity of the distributed values. Why? Because, given your extremely limited knowledge of the distribution, as expressed by a very few constraints, you want to maximize your openness to the real state of the world of which you are quite ignorant. When you choose your diet, you may impose a few constraints on the distribution (e.g. no processed sausages) and maximize the entropy (i.e. your ignorance) of everything else just to keep you on the safe side and minimize your bias. You may decide, for instance, that unless the consumption of meat is morally forbidden, you may consume some meat in order to be on the safe side rather than put all your bets on a purely fruitarian diet.

In his movie *Sleeper* (1973), Woody Allen describes the adventures of a nerdy owner of a health food store who is cryogenically frozen and then defrosted in the future only to find out that smoking is good for your health. Although we know today that smoking is bad, Allen's humorous and highly intelligent movie artistically "explains" to us that in order to be on the safe side, we should always remain open-minded, even to ideas and choices that we currently believe to be utterly foolish. In a deep sense, Laplace, Jaynes, and Allen converge in their basic conceptual understanding, which is quite exciting to realize. What I would like to argue is that through the need to integrate *different* perspectives, a small number of particles may produce a *flexible* and dynamic representation that may keep them open to the uncertainty of the world along the same line of reasoning proposed by Laplace, Jaynes, and Allen. We will illustrate this process in the next section.

Jaynes' idea of maximum entropy may also explain why the Nazis were so wrong in their fantasy of *eugenics*. Let's assume that you are a handsome Nazi officer by the name of Heydrich. You are tall, with white skin, perfect white teeth, blond hair and blue eyes, not to mention a muscular body and musical talent (Wagner is your favorite). After enthusiastically learning about the Führer's ideology, you see no sense in mixing your perfect genes with those of "inferior races" such as Jews. Mixing your genes with short, thin-haired Jews sounds like a very bad idea. However, mixing your genes is like diversifying your investment portfolio under some constraints. No one really knows what the future may bring and putting all your bets on a single blond Nazi girl may be the wrong decision, at least for the long-term future of the collective to which you belong. Imagine a squad of perfectly tall and muscular Nazi soldiers fighting against Montgomery in the hot deserts of North Africa as part of their attempt to establish the Third Reich. In this context, being tall and muscular is probably not an advantage but a disadvantage as the amount of energy required to feed and cool a 90 kg muscular solider in the hot desert is much higher than that required to feed a smaller solider. If there were a limited supply of food, this situation could make a big solider much more fragile than a smaller solider. What shaped the imagined Teutonic ancestors of the Nazi officer in the cold fighting zones of the north doesn't seem to be the best solution for fighting a bunch of "inferior" soldiers, some of them quite small, in the hot deserts of North Africa.

The idea of eugenics is not absolutely foolish. When abortions are carried out in cases of severe genetic defects, they are just constraints imposed to avoid suffering. But taking this idea to its limit results in strategic stupidity not to say moral evil. Irun Cohen (2016) explains why sexual reproduction is so common despite the uncertainty involved in mixing our genes, although he does not use Jaynes' idea of maximum entropy. However, his explanation perfectly resonates with this idea. If you are a Nazi officer who believes that he is the best of the best, then why mix with others at all? Even with a beautiful and pure Nazi girl by the name of Lina, whom you will marry? Why risk half of your genes? The Nazi eugenics illusion edges toward the idea of self-reproduction, and it is totally wrong for the reasons mentioned before.

Maximum entropy is not the whole picture, of course, but an important part of it. When a small number of interacting particles attempt to handle a real-world situation, there are different perspectives that must be communicated and integrated to form adaptive behavior. A small number of bacteria sensing their environment may each sample different portions of the environment. None of them has a bird's-eye view of the environment. Each bacterium has a perspective and each bacterium may communicate this perspective to the nearby others through *local* interactions. Such communications may facilitate the process with a clear benefit to the overall population (Long et al. 2017). The fact that local-to-global processes of communication have been found in such populations indicates that small systems seem to have an inherent benefit, as they present an optimal level between the "freedom" available to the single particle at the price of anarchy, and the saturated state of a big system, where constraints are accommodated to a degree where no "movement" is possible. It also explains why ideas must be formed even by bacteria. As each bacterium holds a limited and particular perspective, its limited perspective cannot serve to form an abstract memory to be used for the future. Some kind of generalization must be produced, and this is only possible by transcending the limited perspective of the individual particle.

The solution is the formation of an idea by somehow comparing the different perspectives and using the topology of this information exchange in order to form an idea in between the members of the group. *As perspectives necessarily change, the topological information formed in between the particles may be changed accordingly, and the collective representation may have the benefit of flexibility for real-time dynamics.* This is my major thesis to explain the context where small systems are at their best: the context of real-time dynamics. Small systems may optimally use ideas and organize the behavior of their particles without falling into the trap of the mob, while at the same time maintaining optimal flexibility as they integrate the *different* and *changing* perspectives of their *individual* members. Perspective-taking, which is crucial for the thesis developed in this book, is further discussed and explained in the next section.

Between Copernicus and the Anarchist

Perspective-taking is how we see a situation from another's point of view. When particles communicate, they don't just send signals but adjust their behavior by observing nearby others. Observation, perspective-taking, and self-regulation are therefore deeply connected. Let me model and illustrate this process through a simple example. A child emigrates from Jamaica to the USA. This child has an idea of what music is all about and this idea can be represented by the words that appear in the context of the word "music." We can learn about this representation if we examine a linguistic corpus and identify the words that appear in the lexical neighborhood of the word "music." We may learn from the Corpus of Global Web-Based English[8] (GloWbE) that, in comparison with the USA, Jamaica has a different distribution of words associated with music. For example, in Jamaica the word "reggae" appears with "music" much more than in the USA (79.53 per million vs. 0.14 per million, respectively) and so does the name "Bob Marley." Let's use an oversimplistic example in which the representation of music by the child from Jamaica may be modeled through the distribution of only three words associated with music (Table 3.1).

---

[8] https://www.english-corpora.org/glowbe.

Table 3.1 The distribution of words collocated with music (Jamaica)

Word	Probability
Reggae	0.7
Marley	0.2
Rock	0.1

We can see that the child's representation of music is expressed by the distribution of words accompanying it. The word "reggae" very frequently appears in the context of "music" but the probability of encountering the word "rock" in the context of "music" is very low. Now, the Jamaican child arrives in the USA and talks with American-born children about music. He (unconsciously) realizes that the American children are talking about music in a totally different way. They seldom mention reggae or Bob Marley, and mainly discuss music in terms of rock and roll. Their representation of music may therefore be modeled by the following distribution (Table 3.2):

How is it possible to model the perspective-taking of the child from Jamaica? One possible way is to use the measure of *relative entropy* or *Kullback–Leibler divergence*. This measure uses two distributions: P and Q. Q is the distribution we hold, and P is the distribution that we would like to understand through Q. In our example, P is the distribution of words used by the American-born children and Q is the distribution that models the Jamaican-born child's idea of music. The relative entropy measure is formulated as follows:

$$D_{KL}(P\|Q) = \sum_{x} P(x) \log \frac{P(x)}{Q(x)}$$

Technically speaking, relative entropy measures the number of *extra bits* of information we need to add in order to code distribution P given our knowledge of distribution Q. Without getting into too much detail, and from a simple perspective, the relative entropy actually measures the effort required to adjust our prior representation (Q) to the new representation (P). In our case, it is a measure of how much effort the child from Jamaica must invest in order to take the perspective of his new American friends. Let's illustrate

Table 3.2 The distribution of words collocated with music (USA)

Word	Probability
Reggae	0.1
Marley	0.1
Rock	0.8

Table 3.3 Comparison of the two distributions

Word	Q	P
Reggae	0.7	0.1
Marley	0.2	0.1
Rock	0.1	0.8

how this works by tagging the Jamaican child's representation as Q and the representation of music by the American children as P (Table 3.3).

The next step is to calculate the relative entropy score as follows:

$$Relative\ entropy = 0.1 \times \log(0.1/0.7) + 0.1 \times \log(0.1/0.2) + 0.8$$
$$\times \log(0.8/0.1) = -0.28 + -0.1 + 2.4 = 2.02$$

There is no absolute meaning to the number 2.02. The relative entropy is 0 when no effort has to be made to adjust Q to P, and when the relative entropy score is higher, it quantifies the price that we have to pay in order to adjust our representation. In our case, the number puts a price tag on the effort required by the child from Jamaica to adjust his perspective to that of his new American friends, or to take their perspective and see music as they do.

What about the price to be paid by the American children if they would like to take the perspective of their new Jamaican friend? The price is *asymmetric* as the measure of relative entropy is *by definition* asymmetric. This is what makes it so relevant to modeling perspective-taking. The effort required by the child from Jamaica is not the same as the effort required by his friends. In this case the American children's price is lower: 1.86. To adjust their perspective to the perspective of the child from Jamaica, the American born children would pay a lower energetic price.

We have shown a way in which representations can be modeled by distributions and perspective-taking through the measure of *relative entropy*. We have also shown that relative entropy is a good measure for modeling the effort required to take the perspective of others when they hold a different representation of the world. The next step is to explain how constraints imposed by multiple perspectives may lead to an abstract representation that can be flexibly used by a small social system of particles holding and communicating these different perspectives.

We are now getting into the main thesis of this book, but first we have to discuss George Zipf's (2016) *principle of least effort*. Zipf proposed the reasonable hypothesis that cognitive systems constrained by limited energy resources are motivated to save their efforts when operating in the world and communicating with others. In the above context, we may imagine a situation where the Jamaican child and his American friends are trying to form a general and shared idea of what music is all about. They hold different perspectives, as expressed by their own different distributions. Moreover, each of them is obliged to pay a different price to take the perspective of the other and this price of perspective-taking can be modeled using relative entropy. The price for the child from Jamaica is 2.02 extra bits of information and the price for the American-born children is 1.86 extra bits of information. When trying to form a general idea of what music is all about, each has to take the required effort of the other side into account. It may be that neither the Jamaican child nor the American children want to invest the full amount of energy in adjusting their perspective to that of the other. However, the price of anarchy might be that they would not have a more general and abstract idea of what music is. How is it possible to build a new distribution under the constraints forced by the non-converging perspectives and their entailed energetic prices?

"Show Me Yours and I'll Show You Mine"

The solution may come from a strategy that we may jokingly describe as "show me yours and I'll show you mine." The idea is that as each of us is energetically unmotivated to fully adjust our representation, we may imagine a stepwise process of *mutual disclosure* that may benefit both interlocutors. Let's try to introduce a heuristic to address this challenge and its consequences by using another example.

Two people are observing a situation that may be represented using a vector of four features such as the components of their diet. The vector is an array of numbers or probabilities that sum to 1. In our case, the vector is that of diet, and the ideal distribution of food as each person conceives it. Each of the people holds a different perspective, which is reflected by the different distributions. We are given two vectors or perspectives and agree to cooperate in communicating them through a mutual process of stepwise disclosure. We can think about the values in each cell of the vector as coupons. Each of us gets "coupons" of energy, which sum to 100%. Here are the vectors (Table 3.4):

We start with person A and feature 1. Why feature 1? Because it's A's highest bet. Taking his perspective, I may believe that my heuristic is that I will start from my highest bet, which represents my strongest belief, and will then move down to the lowest.

Table 3.4 The distribution of the diet's features

Feature	Person A	Person B
1 (grains)	90	1
2 (meat)	1	53
3 (vegetables)	4	45
4 (fruits)	5	1

When I'm trying, as person A, to guess the perspective of person B, I hypothesize that feature 1 is represented in her mind with probability of 0.90, as it exists in my own mind. By default, this hypothesis is the cheapest as I don't have to invest energy in changing my perspective. However, under the assumption of full ignorance (i.e. *Laplace's principle of indifference*) of person B's own representation, my guess should be that the value to be found in cell 1 of person B's distribution is 25. Why? Because if you have 100 coupons and the distribution is homogenous across the four cells of the distribution, then you should find 25 coupons in each cell.

At this point, we see that, as person A, I am in conflict. When I attempt to guess the number in cell 1 of B's distribution, I am unmotivated to change my mind and under the *principle of least effort* may use my own (limited) knowledge. In this case, my guess is that the value of cell 1 in B's distribution is 90. This is a kind of egocentric thinking that you may sometimes see among young children and (unsurprisingly) among some adults as well. The first motivation is therefore to stick to your own knowledge and limited perspective as it is the energetically cheapest move. However, such a limited perspective might be dangerous as your perspective is by definition limited. This is such a commonsensical idea that it requires no deep theoretical explanation. In fact, human culture has provided several derogatory ways to describe people who are so self-centered that they cannot see beyond their own limited perspective. The benefit entailed in saving effort by "having your head stuck up your ass" is therefore in conflict with the price of a limited perspective that might miss something important.

As you have no idea about what exactly is the other's representation, the *principle of indifference* suggests that your ignorance should be expressed by maximizing the entropy of the other's representation. Here comes the *second conflict*. On the one hand, you would like to save your effort and to stick to your own knowledge and perspective. On the other hand, given your interaction, you are asked to maximize the ignorance of your expectations. How can you resolve this conflict?

A very reasonable and commonsensical heuristic is simply to find the *middle point* between your own limited perspective and knowledge, and your hypothesis under full ignorance of the other's perspective. A better guess may therefore be to somehow average the two guesses: my original hypothesis and the one formed under full ignorance. I use the average of 90 and 25 and the result is 58 (rounded).

Now the game I play with the other is a game of disclosure. After I have guessed what hides in B's cell, she plays her role in the game and shows me the value of her cell. When I'm exposed to the representation of B (let's now call her Penny, hearkening back to the *P* of the previous section and let's call me Quentin), I realize that feature 1 holds only 1%of her distribution. At this point, I understand that there is a difference between our perspectives. While I originally believed that grains would be 90% of her diet and under the principle of indifference corrected my "bet" to 58%, I have now found out that Penny believes that grains should only make up 1% of her diet. I understand that I'm wrong but I also learn that in Penny's distribution there are 99% of the coupons left and that I have to guess their distribution. This is an important constraint. I also know that 58 coupons have been erased from my own account and that I am left with $100 - 58 = 42$ of the coupons. Therefore, I adapt my cells accordingly:

Feature	Quentin	Penny
1 (grains)	58	1
2 (meat)	1/10*42 = 4	53
3 (vegetables)	4/10*42 = 17	45
4 (fruits)	5/10*42 = 21	1

I now move on to guess feature 3 (Fruits), which is the second-ranked feature in my distribution. Under the assumption of ignorance, I guess that the value of feature 4 in Penny's distribution is 33 (i.e. 99/3). My updated guess (i.e. 21) is now adjusted to become (0.5 + 0.33)/2 *21 = 9:

Feature	Quentin	Penny
1 (grains)	58	1
2 (meat)		53
3 (vegetables)		45
4 (fruits)	9	1

I continue with this process so that I end up with a new updated vector:

Feature	Quentin	Penny
1	90	58
2	1	16
3	4	17
4	5	9

The result of this process is the entropy of my new distribution is $H =$ 1.63, which is higher than the entropy of my original distribution (i.e. $H =$ 0.60). This means that the interaction with Penny has led me to increase the entropy of my representation under several natural constraints:

(1) the principle of least effort,
(2) the principle of indifference, and
(3) Penny's own perspective on the world.

This is an extremely important point. The entropy of my representation has increased under *natural constraints forced through interaction with another*. We are not talking about maximizing entropy as a kind of optimization problem that requires the use of sophisticated mathematical tools such as *Lagrange multipliers*. We are talking about *increasing* (rather than maximizing) the entropy of my representation by using interaction and natural constraints: my basic selfish interest, my basic ignorance of the world, and the need to attune to another's perspective through a mutual process of disclosure.

To recall Herbert Simon's idea of *bounded rationality* (Simon, 1957), under the cognitive limitations in which we operate, we tend to seek a *satisfactory* solution rather than an *optimal* solution. In other words, given our limited resources, we seek good-enough solutions rather than the best solutions. The idea of bounded rationality has repeatedly popped into my mind when I have been cleaning my family's house. When cleaning the house, I have realized how difficult it is to reach an optimal solution that is close to that of a professional. Such an "ideal" solution is extremely demanding. I have chosen a satisfactory solution not through laziness but by relying on great minds such as Simon.

In the above example, I (as Quentin) could have considered the problem as an optimization problem, and could have used my mathematical knowledge to solve it through Lagrange multipliers, but, even for a scientist modeling the above situation, it is far from trivial to decide how to determine and quantify the constraints to be included in the model. Believe me, I have tried this approach with some highly intelligent friends who hold PhDs in physics and mathematics. Moreover, it seems problematic to assume that perspective-taking in the natural world has been applied through mathematical optimization. Have all beings—from bacteria to human beings—been equipped by nature to unconsciously calculate Lagrange multipliers? The above procedure presents an extremely simple, natural, intuitive, and cheap heuristic for increasing the entropy of my own representation through interaction with another.

I have described a process in which only one party plays the game by guessing, but we can imagine a mutual and simultaneous process working according to the logic of "show me yours and I'll show you mine." In this case, we reach even better results. What we can see is that through a stepwise process of perspective-taking, as modeled according to one possible heuristic, both parties may generate representations that are closer to their interlocutors and that have a higher level of entropy—not maximum entropy under constraints but higher entropy under constraints. This is bounded rationality as compared with an optimal solution. If I must bet on the representation of the thing each of us observed, then the new vector seems a better choice than the original one.

The *Copernican principle*, in its general form, states that as observers, we don't occupy a special place in the universe. The uniqueness of our place results only from the unique perspective we have on the world. As beautifully stated, "We are all unique but we are never alone" (Holquist & Liapunov in Bakhtin, 1990, p. xxvi). The perspective-taking heuristic, as represented above, is a way to reconcile two allegedly opposing forces. On the one hand, each of us holds a unique perspective. However, on the other, because according to the Copernican principle we don't occupy a special place, perspective-taking potentially allows us to reciprocally build a better representation of the world. This process comes with a limit and some necessary qualifications. As the number of particles in the system increases, constraints are accommodated, our representation converges to the mean, a symmetric structure is formed, and we start to lose the flexibility of a small system. The *wisdom of crowds* is wisdom limited only to certain situations, such as those where we seek to estimate a group average. Now, do you remember that I promised to explain why playing Russian roulette is not such a good idea?

On Russian Roulette and Misleading Averages

The first time I learned about Russian roulette was when as a high-school student I watched *The Deer Hunter* (1978). This powerful film, featuring a wonderful set of actors from Robert De Niro to Meryl Streep and Christopher Walken, includes two shocking scenes involving the playing of Russian roulette in Vietnam.

Russian roulette is a simple "game": it involves a handgun with six chambers, a bullet inserted into one of them, and a strong rotation of the cassette that should randomly turn it, after which the player turns the gun on himself and pulls the trigger. The chance of being killed is one in six, or 0.17. The chance of staying alive is 0.83. Let's imagine that during the sixteenth century, a group of 100 peasants are asked by their decadent Russian landlord to play the game. They are offered a deal. Each peasant plays once. The payoff for survival is freedom from serfdom, a bottle of fine vodka, and 100 rubles. When the gun shoots a bullet, the price requires no further explanation.

While morally horrible, the proposal may not be such a bad idea for the collective. Most of the people will survive, will be free, and will gain the offered benefits. The price of those who would not survive is a price that might be paid in any case if they considered rebelling against their serfdom. On the group's level of analysis, the average may be a good measure to be considered in their decision-making process.

Now imagine a different scenario where the peasants are shocked by this proposal and refuse to accept it, except for Ivan the fool. "I'm ready to play the game instead of you," he announces, "and will pull the trigger 100 times for the potential benefit. After all," he adds, "the chance of being saved (i.e. 83%) is much higher than the chance of getting a bullet in my head. Isn't it a good bet?" It is not, as poor Ivan may end his life shortly after the game starts. The average that is relevant to a group at a specific time-slice is not necessarily the same as that for an individual repeatedly playing the game. Now imagine a whole group of people trying to adopt Ivan's line of reasoning by volunteering to play the game over and over again… Playing in the real, non-ergodic world with its irreversible and fluctuating nature requires some different strategies—averaging does not always work.

When a small group of people interact and form a flexible representation that can adaptively change and that is attuned for real-time decision-making, they do not strive to estimate the mean of some population, a process limited to certain contexts only. They may naturally be led through a limited number of mutual constraints into a representation that may better use them in order to address the unknown future. Averaging is not an optimal strategy everywhere, otherwise the averaging of our genes would have become common in nature. The average, or the arithmetic mean, is just one possible *loss function* that we may use to describe a population. It is not the ultimate solution to every challenge that exists in the world. While the wisdom of crowds may be formed in large systems where it is important to balance deviations, small systems are not as interested in averaging as in creating online solutions in real time. Moreover, previously I proposed Jaynes' principle of maximum entropy in order to explain the benefit of a group. However, to model real systems, it may be even better to adopt Ashby's (1991) idea of *requisite variety*, which is discussed in the next section.

Indiana Jones and the Law of Requisite Variety

Ashby's *law of requisite variety* is presented in the context of regulating the behavior of a system in relation to a set of disturbances. Imagine, for instance, a soccer team striving to move the ball forward in order to get closer to its opponent's end of the pitch in order to score a goal. The team is generally moving the ball forward, but encounters disturbances formed by the other team. The other team has no interest in paving the way for its opponent and therefore its members try to block and hijack the ball when it is passed. In this context, the variety discussed by Ashby is the variety of responses that aim to regulate the disturbances to the extent to which they don't prevent the system from achieving its goals. To repeat, Ashby's law of requisite variety is relevant to contexts where a system is striving to regulate its behavior in the face of external or internal disturbances. Regulation is expressed through a set of responses and this set has a variety indicating the available repertoire. The variety, Ashby argues, must meet a certain level in order to address the target of regulating the behavior and overcoming the obstacle formed by the disturbances. He gives the example of a fencer facing an opponent who has various modes of attack. To address these attacks, the fencer must have a rich-enough repertoire of attacks, defenses, and counterattacks.

Do you remember the fictive adventurist and archeologist Dr. Indiana Jones? In the film *Raiders of the Lost Ark* (1981), there is a memorable scene where Jones is facing an expert Arab swordsman in the streets of Cairo. The swordsman is a huge, scary guy with a vicious appearance. When he confronts Jones, he pulls out his huge sword and does some tricks showing his mastery of the weapon. His expectation is that Jones will challenge him with the only "weapon" that Jones holds in his hand, which is… a whip. Jones, however, casts a "cool" look at the swordsman, pulls out a gun, and shoots him dead. Hilarious!

The swordsman's threat was a potential disturbance to Jones, but Jones' variety of responses regulated this situation in a perfect and unexpected way.

He has exceptionally flexible mind and the ability to improvise in the most difficult situations. Jones did not expect the unfair challenge of fighting a swordsman armed only with a whip. However, his repertoire of responses included something that the vicious swordsman could not have expected. The requisite variety for regulating internal or external disturbances must therefore meet a certain threshold according to which no disturbance will be able to overcome the system's behavior. The variety available to address disturbances must be big enough to cover the whole spectrum of possible disturbances. In practice, though, we understand that this variety can never be enough. We must acknowledge that (1) the set of responses is finite, (2) the distribution of responses should somehow correspond with the real distribution of disturbances, and (3) there should be some kind of bootstrapping procedure through which the original set of responses may be flexibly adjusted and changed.

I'm always explaining to my students that if they want to really understand the law of requisite variety then they should examine the evolution of mixed martial arts (i.e. MMA) from the early days of the UFC (Ultimate Fighting Championship). The first championship took place in 1993, and martial artists were called to test the virtue of their martial art in vivo against opponents from different disciplines. As you may know, the martial arts include many disciplines and schools, and the effectiveness of some has been hotly debated. For instance, some martial artists have argued that they can beat their opponents using a *death touch*, as popularized in Tarantino's *Kill Bill: Volume 2* (2004). The UFC aimed to test these pretensions against the solid rock of an opponent's knuckles.

The first fighters in the UFC arrived from very specific disciplines of the martial arts: from taekwondo, which is a Korean form of martial art specializing in impressive kicking, to boxing and sumo. Each fighter's variety of responses was largely limited by the discipline from which they originated: boxers had expertise in punching but were unexperienced in kicking and grappling, taekwondo guys were proficient in kicking but limited in grappling and striking, and grapplers were poor strikers and kickers. Those with a limited repertoire were usually easy prey for their challenging opponents. When I watch some of the early fights of the UFC, I cannot avoid thinking about a poor panda thrown out of its bamboo forest into a branch of McDonald's and trying to feed itself by eating something its evolution has not prepared it to digest. The evolution of the UFC was such that "panda fighters" simply did not survive. The requisite variety of their responses was generally too limited to handle the tricks of challenging opponents. Today, you would not see a panda fighter in the UFC arena. Nor sumo fighters, masters of internal chi, or boxers who have no idea what to do when they find themselves on the ground. Today, a UFC fighter has the requisite variety to address various modes of attack, from striking to kicking and grappling. However, while requisite variety is a must, some decisions have to be made in order to prioritize certain responses. When preparing for a fight with an excellent striker, there is no point investing the same amount of time in practicing strikes and roundhouse kicks. The diversity of the fighter's portfolio must be carefully managed. In addition, the variety should not be a closed set. When preparing for a fight with Jon "Bones" Jones, one must increase one's variety to included potential responses to some creative moves this fighter is known for, such as a back-spin elbow. When we operate in a small group, the balance between the creativity of our individual anarchy and the constraints forced by the members may provide the optimal context to produce the requisite variety for regulating the system's behavior in order to address its goal.

Weaving the Threads: How Tailor-Made Solutions Are Formed

We have made a long journey since the first chapter, where the case study of Leicester was presented. Let me summarize the ideas presented so far. The most basic cognitive activity is the recognition of distinctions (e.g. entropy states) and the association of these distinctions in a way that forms an informative pattern. To be able to conduct this activity, we must learn some patterns, where structural associations are formed, and order patterns, where the temporal appearance of events is understood. In a fully irreversible world, where it is impossible to go back in time and to restore the inputs from an output, it is impossible to learn such patterns unless we are equipped with a memory device and some way of generating abstract representations (i.e. ideas) that can sustain the entropic force of nature. A memory device along the lines of reversible computing may be expensive to maintain as the numbers of inputs and the outputs are the same. Overloading our system with memory traces is therefore not a feasible solution. It seems that some information must be erased and cleared from the system in order to form abstract representations on a higher level of analysis. In order to represent the world and memorize, we must be able to forget—to throw away some information and distinctions in favor of more abstract distinctions. This process is performed when information is processed in-between particles and some of it is lost in favor of a more general representation. When a group computes, some information is lost as a result of the computation process. As a result, abstract representations (i.e. ideas) are formed and stored in between the interacting particles. When we collectively digest information, we give up some information on the micro level in order to generate new and more abstract structures (i.e. ideas) on the mesoscopic and macro levels of the system. Small systems such as families and soccer teams are unique in the sense that each particle is an autonomous computational engine that represents its world in a way that reflects both its own unique perspective and the perspectives shared by others. Order is formed in the mind of the collective when abstract patterns are used to monitor and regulate its behavior.

Moreover, when interacting, the particles impose constraints on the behavior of each other. There are two major sources of constraints that we have discussed so far:

(1) the need to see through another's perspective, which explains why symbolic activity mediates interaction, and
(2) the need to resolve conflicting interest, such as that exposed by the principle of least effort.

These constraints are crucial for understanding the behavior of small social systems. When we take into account these two constraints, the emerging product is a form of collective representation or idea that is much more abstract and flexible than the local representations held by each particle and more flexible than the representation formed and maintained by a large collective of individuals. The representation formed by a small system is abstract because information is necessarily lost as we move from the particle's level of analysis to the system's level of analysis. It is more flexible because it maximizes the entropy/requisite variety under constraints of perspective-taking and the principle of least efforts, as I have shown before, and it is adaptive because when perspectives change, they have the potential to change the system's representation and behavior accordingly.

A soccer team has the objective of scoring a goal. Regulating its behavior under the disturbances forced by its opponent and by its internal fluctuations leads to a form of flexible and dynamic model of itself and its opponent. As explained by Conant and Ashby (1970), every good regulator of a system must be a model of that system. The model formed by the soccer team cannot be in the form of a stable and limited religious dogma or even a scientific model. It must be a model or representation that regulates the real-time and ongoing behavior of the team against its opponent. A soccer team with the mentality of a religious sect is destined to fail. The real potential merit of a small system is therefore to form tailor-made solutions to real-time situations by converging perspectives and merging the relatively free behavior of the individual particles with the more ordered behavior of the system, which given its size and location in a real-world situation is not yet subject to the paralyzing behavior of a larger collective.

Small groups therefore form a level of analysis that has great potential for flexible forms of adaptation. Small systems are not subject to too many constraints, at which point a system becomes saturated and too stable. They potentially inhabit an intermediate level where the "anarchy factor" is kept while being ordered enough to form a working collective. This point can be illustrated through music and a comparison of a symphonic orchestra with a jazz group.

A jazz group is composed of a limited number of individuals. It can be a trio (e.g. the Bill Evans Trio), a quartet (e.g. the Dave Brubeck Quartet), or a quintet (e.g. The Jazz Messengers). In contrast, a symphonic orchestra, such as the Berlin Philharmonic, may include more than 50 players. Although there are trios that play classical music, they don't improvise, and you won't easily find a band of 100 players improvising as a collective. Improvisation is what jazz is all about, and it has deep roots in the "dialogic"[9] communicative sources of this genre. The free spirit of jazz is expressed in improvisation. Improvisation doesn't mean that anything goes and that you play only with yourself in an idiosyncratic manner. *Improvisation is maximal freedom under dialogic constraints*. It is a process that expresses the individuality of each player within a small collective. When you hear Chick Corea playing with his trio,[10] you can hear dialogic music at its best and will intuitively understand what I'm talking about. You will hear improvisation on known pieces and themes such as those produced by the brilliance of Bach. This is not a performance by protocol, as can be seen in the case of symphonic orchestras playing classical pieces of music (e.g. *The Passion* by Bach), but a dialogic performance that maintains a delicate balance between the *individuality of the players and their own and real-time interpretation within the composite system of the trio as a whole*. A jazz trio is the best example I can think of to illustrate the balance between the anarchistic and individual perspective, with its benefits and shortcomings, and the collective activity, with its own benefits

---

[9] I put the term dialogic in quote marks as it has been severally abused by various post-modernist academics, who have subordinated it to the politics of identity.

[10] See, for example, https://www.youtube.com/watch?v=KRLwsSbqlwo&list=WL&index=21&t=0s.

and shortcomings. I can barely imagine the possibility of interpretation and individuality in the Berlin Philharmonic.[11]

Small systems therefore have the benefit of small scale, where collective activity involves the representation of the world through various perspectives and under the selfish interest of investing minimal effort in communication. These constraints and dynamics push the solution formed by the collective into an abstract (but not too abstract), flexible, and dynamic representation through which the system may adequately respond to real-world challenges in real time. If you are a parent caring for his children and looking for real-time solutions, then you probably know how important the family is in providing creative solutions. These solutions are not creative in the sense of uniqueness. A mother or father who finds a way to calm a crying and anxious baby usually feels as if s(he) has been involved in a creative act in the style of Michelangelo. Through their synergistic activity, the interacting particles may form new solutions to real-time challenges, a process that has been described as "autocatalytic" (Atlan & Cohen, 1998; Cohen, 2016). Responding to a challenging move by the opponent team, addressing a baby's concerns by working as a dyad of parents, and improvising by listening and conversing with the other players in a jazz trio are all forms of synergistic creativity observed in small social systems. Saying that, I do not mean that small systems are necessarily optimal or effective. The components may lack "chemistry" or simply behave in an additive manner, perspective-taking may be missing under the tyranny of a single or a few particles, and relations may be symmetric to the extent that the topological information of the system produces no added value. Small systems have the potential to work extremely well in certain situations but, as we all know, such potential may not be fulfilled given certain circumstances. Having hopefully gained a broad and deep understanding of some possible theoretical scientific foundations for understanding small systems, we may now turn to the way different aspects of their behavior may be modeled. In the next chapters, we will learn about Thai boxing and love triangles; chemistry between soccer players and within romantic couples; why it is important for both families and soccer teams to be messy enough; and what lesson the Brazilian football team could have learned from an Antarctic expedition.

---

[11] It goes without saying that this discussion doesn't mean that jazz is "better" than classical music or that trios are better than orchestras. Such a childish and naïve misunderstanding should be immediately rejected.

Summary

- Maxwell's demon teaches us that an information-processing capacity underlies every living system.
- At the heart of living systems there is a process of computation that is grounded in abstraction.
- We learn abstraction through irreversibility forcing constraints on the behavior of a system.
- However, to abstract, we must go back in time, which is impossible in an irreversible world.
- The solution is memory, which is the use of traces pointing to past events.
- Memorization is a process necessary for learning patterns in an irreversible world, specifically through some form of collective activity that generates abstract representations through a social system of signification.
- Small systems are good at generating abstractions because a lot of information exists in between the individuals of which they consist.
- The flexibility of the collective representations is formed through limited individual perspectives integrated with the perspectives of others.
- As perspectives necessarily change, the topological information formed in between the particles may be changed accordingly, and the collective representation may have the benefit of flexibility for real-time dynamics.

Part II

Understanding Small Social Systems

4

The Enemy of Art Is the Enemy of Soccer: On Constraints, Thai Boxing, and Love Triangles

In this chapter, we learn what art and soccer have in common, what Thai and Burmese boxing can teach us about constraints, what a famous Spanish painting can teach us about perspectives and triads, and why constraints are so important for our understanding of small social systems.

What Art and Soccer Have in Common?

A saying attributed to the famous film director Orson Welles is that "the enemy of art is the absence of limitations" (Jaglom, 1992, p. 78). This sounds like a paradoxical statement as art is sometimes conceived to be the field expressing ultimate freedom, which should be beyond social norms, censorship, or any other form of constraint. Art, according to this rather naïve conception, is the expression of imagination and creativity and as such must be *limitless*. There is a point to this idea as we are familiar with the poverty of ideological art as produced, for instance, in the former Soviet Union. This form of recruited art aimed to serve the revolution and operated under severe constraints. By playing the role of a solider fighting for the revolution, the Soviet-recruited art lost its free spirit, which seems to be the essence of great art. However, if you give the idea a second thought, then you realize things are not so simple and that the imaginary and absolute form of freedom propagated by some young students of art is the ultimate form of slavery, as nothing significant can be created without constraints. Some of the greatest

Y. Neuman, *How Small Social Systems Work*, The Frontiers Collection,
https://doi.org/10.1007/978-3-030-82238-5_4

pieces of art, such as those produced by Bach (e.g. "Prelude and Fugue No. 1") and Michelangelo (e.g. the Sistine Chapel ceiling), were created in the context of the subordination of the artist to more or less fastidious patrons and constraints of form and content. In these cases, the artists exercised their creativity under limitations. From a naïve perspective, one might barely realize how visual art and music have flourished under the dogmatic Catholic Church and how disappointing are the outcomes of modern visual art in wealthy and open post-modernist societies, where the freedom and welfare experienced by artists have no precedence in history.

So far in this first part of the book, I have introduced and discussed the idea of constraints in the context of optimization. In mathematics and computer science, a constraint is a condition of an optimization problem that the solution must satisfy. For example, let's say that Abe has just bought a new car. As a highly neurotic person, he starts worrying immediately and asks himself when exactly he should sell the car in the future. In other words, he is interested in determining the best time to sell the car in the future. The context is simple: he has bought a new car and, as the value of the car declines over time, he would like to know *when* to sell it in order to minimize his losses. The optimization problem therefore asks when Abe should sell the car in order to minimize his losses: should he sell the car after one year? Two years? Five years? And so on. The car's market price is of course not the only variable that Abe should take into consideration. In this deceptively simple optimization problem, there are other factors to be taken into account, such as Abe's personal joy in the car, the cost of fixing it, common car issues (e.g. a sputtering engine), the deprecation of this specific model of car, and so on. Having put all of these variables into the equation, we can add some *limits to the solution space* of the problem by saying, for instance, that Abe would like to find the best price that is higher than 50% and lower than 80% of the car's original price. This context of optimization has mostly been used in this book in analogical or metaphorical terms, as there are difficulties in applying the idea of optimization to small systems in a straightforward manner. Indeed, a soccer team may want to optimize its goals balance and a family may want to optimize its happiness, but such a form of rationality imposed on the behavior of a small system is naïve and of limited explanatory power. The constraints that I discuss in the context of small systems are not boundaries directly and explicitly imposed on the solution space of a well-formulated optimization problem but limits naturally emerging from interactions that shape the behavioral space of the system and its unique nature in a bottom-up manner.

In many natural systems, constraints function as a form of "downward causation" through which "boundary conditions" existing at the system's level of analysis channel the behavior of the system at a lower level of analysis (Noble, 2012). The systems biology approach presented by Dennis Noble (2008) urges us to think about constraints as "boundary conditions" through which the behavior of the system's components is channeled. Constraints may emerge through bottom-up interactions, but, given their formation through micro-interactions, they start functioning as boundary conditions operating on the same micro-level units that generated them. This natural approach to constraints is the one adopted in the current book.

In the real-world examples that I have used so far, we were not interested in formally and consciously optimizing a solution to a problem but in modeling real-world situations where under the idea of *bounded rationality* (Simon, 1957), the particles may sometimes change their solution space, which means flexibly adjusting it to a real-time situation. Improvement doesn't necessarily accompany the interactions of a small social system, specifically when examined in real time. However, if there is an improvement then this *improvement*, rather than optimization, will naturally emerge from interactions under some "boundary conditions," such as minimizing energy and the need to be involved in perspective-taking. To better understand such real-world constraints, which in themselves may dynamically change, we may examine the difference between Thai boxing—*Muay Thai*—and its less "civilized" relative, the form of Burmese boxing known as *Lethwei*.

What Thai and Burmese Boxing Can Teach Us About Constraints

Muay Thai is a fighting sport in which two fighters may use their fists, elbows, legs, shins, and knees to beat their opponent. It is a brutal form of fighting sport, and bloody injuries are more common than in other such sports, such as judo or wrestling. Nevertheless, in the past, Thai boxing was even more brutal, but, similarly to modern boxing, it has been restrained in order to adjust it to the modern world. The encounter between ancient forms of Thai boxing and the English style of boxing, with its codified norms of behavior, led to the modern style of Muay Thai as we know it today. This modern style includes the use of gloves and a scoring method applied by judges. The judges score the performance of each fighter throughout the fight using criteria such as aggression and effective technical execution. A roundhouse kick to your opponent's head will score higher than a simple punch. This scoring method significantly influences the strategy of the fighters and the way they allocate their energetic resources across the rounds. For example, in the first round each fighter usually examines his opponent with caution.

In contrast with Muay Thai, the Burmese style of boxing—*Lethwei*—is a less civilized and much more brutal kind of sport, as the fighters don't use gloves and are allowed to use headbutts (!), and there is no scoring system. A fight can end in one of only two ways: knockout or draw. The major motivation of each fighter is therefore to knock out his opponent rather than to score more points.

The use of gloves in Muay Thai is an imposed constraint that has some interesting implications. Not only it is a way of reducing severe injuries (which are considered to be inappropriate for a gentlemen's sport) but it also indirectly increases the repertoire of the fighter and the complexity and potential interest of the fight. It may sound paradoxical but introducing the very simple constraint of gloves probably led to the increase in this sport's complexity. The reason is that gloves allow a fighter to cover a larger part of their head and to more easily absorb strikes. As such, the fighter naturally uses more kicks and "fancier" and riskier techniques than one may observe in a bareknuckle fight, where each move might expose the fighter's soft spot. From a "romantic" and naïve perspective, a bareknuckle fight, such as those described by Ernest Hemingway (1970/2013) or practiced in the UK for generations, might be perceived as much more fascinating than the "codified" form of Thai boxing. However, if you have ever observed and compared Muay Thai and bareknuckle fights, you must admit that the civilized form is richer and much more interesting in several senses. Paradoxically, by constraining some aspects of the fight, the complexity and interest of the fight as a whole significantly improves.

This paradoxical aspect is highly important to our understanding of constraints as it exposes a gap between our naïve understanding and the complex reality. Take, for example, what is known as the *Braess paradox*. "Braess' paradox is the observation that adding one or more roads to a road network can slow down, rather than speed up, overall traffic flow through it."[1] It is a "paradox" because we naïvely expect that adding roads would contribute to the flow of traffic and by no means would slow it down. However, as is nicely explained on a popular scientific platform, "The addition of options is not necessarily a good thing."[2]

Game theory shows us that limiting the freedom of *individuals* may sometimes increase their benefit as a *whole*. The reason is that sometimes

---

[1] https://en.wikipedia.org/wiki/Braess%27s_paradox.

[2] https://brilliant.org/wiki/braess-paradox.

accessing a new opportunity is an option so appealing to the "players" that everyone uses it and counterintuitively hurts themselves as a collective and as *individuals within the collective*. Adding options is not necessarily a good idea and adding constraints is not necessarily a bad idea. Once we understand this point, we become fully exposed to the complexity of small social systems and can start asking questions, such as: How may the addition of constraints *nonlinearly and non-trivially* change the behavior of the system? Which constraints are beneficial to the whole? What is the limit line of the added constraints? Is it the point where the system becomes saturated? How constraints shape the form of interactions? And so on. When the objective is to find the right mixture in a soccer team, these questions are extremely important. Finding the best mixture of players is far from a trivial task. Therefore, asking the above questions may not only increase our knowledge but also, and of no less importance, increase our awareness of our ignorance and the way in which we may resolve it.

In our boxing context of constraints, adding constraints (e.g. wearing gloves in a fight) is like removing some roads from a network and lowering the system's degrees of freedom. Is there some kind of reverse Braess paradox where removing a road *increases* the flow of traffic? The answer is positive. In a paper described in a popular science article,[3] it was found that closing some roads and removing some options prevents individuals from adopting the *best selfish solution*. In other words, there is some justification—and from more than one point of view (e.g. game theory)—to the idea that adding constraints may improve the behavior of a system as a whole and will benefit each of its selfish particles. If you remember Zipf's principle of least effort (see Chap. 3), then you may also understand why the imposed constraint of communicating with others while minimizing one's effort has pushed human language to become a complex communication system—one that involves *ambiguity*, for instance (Fortuny & Corominas-Murtra, 2013).

In the context of Muay Thai, we can see that adding constraints reduces the potential risk of some techniques that may have great appeal to a selfish fighter. After all, hitting your opponent with bare knuckles or a headbutt seems to be a much more powerful way of ending a fight. However, as *both* fighters may selfishly use the "road" of bare knuckles and headbutts, any selfish advancement would be lost within the whole system and the result, in terms of the collective of fighters, would be suboptimal. The result of fighters wearing gloves has been a significant increase in the *variety* of techniques used and the *volume* of their use in fights, an implication that has

---

[3] https://www.scientificamerican.com/article/removing-roads-and-traffic-lights.

probably increased the popularity of this sport. Constraints are highly important for understanding small social systems and they oblige us to think in a negative rather than a positive way, which is quite challenging. The meaning of thinking negatively is to ask how certain ways of *reducing* the particles' degrees of freedom changes the behavior of the system and in what sense.

The constraints on Muay Thai have been imposed from the top down, but constraints can naturally emerge through bottom-up processes (i.e. numerous cycles of interaction between players), eventually leading to new norms and codified behavior. We may better understand this point by using the simpler case of dogs' play.

When dogs play, they use a limited repertoire of moves that mostly involve chasing and biting. However, in contrast with a real fight, they don't bite each other to the point of causing death or even significant harm. They play in a civilized and codified manner without a Thai king to impose any rules on the fight. Codified behavior among dogs developed in a bottom-up manner. The aforementioned Gregory Bateson (1972/2000), a brilliant polymath who was interested in such situations of play, provides us with worthy insights about the way such situations should be approached. He realized many years ago how important it is to analyze a situation such as dogs' play (or in our case a Muay Thai match) by paying close attention to the fact that such a system is always embedded within a larger system, and by paying close attention to the way information flows between the players and between the different levels of the system. Today, the relatively new field of multilayered networks (Aleta & Moreno, 2019) is moving us closer to the approach Bateson proposed as early as the 1960s, in the sense of modeling a system as multilayered.

Constraints are about "blocking some roads," with the idea being that the whole system may benefit from this move or at least may change its behavior in some consistent manner. I can play my guitar in my basement, making no effort to attune myself to other players. However, the richness of a jam session played by an ensemble cannot be gained without paying a price. Getting into the groove requires me to accept some constraints imposed through interaction with others.

As already discussed, constraints quickly accommodate as a function of how many players participate in the game, and at a certain point, when there are too many players, the accommodated constraints can move the system into a new phase space where the prior benefits are washed off. At a certain point, it seems that the size of the system does not support the flow of information that is necessary for improvisation, and the system is managed in different ways, mainly through hierarchy or convergence to a Gaussian distribution, where minor random differences between the particles cancel each other out to produce an averaged behavior.

I must emphasize that accommodated constraints are not necessarily a bad thing. They may simply mean that a system is self-organized and has become more ordered in its behavior, which is a wonderful thing for a system of particles whose collective behavior requires a high form of coordination and optimization. However, the price might be that the contribution of the individuals is suboptimal, which in *some contexts* might be detrimental to the functioning of the system. Improvising with too many players seems to destroy a precious aspect of jazz and increasing the number of soccer players on the field to 20 or more per team means that no joy can be gained from the game.

Here is another example, showing how the benefit of a small system cannot be improved by scaling it in size. Nuclear families have existed for millennia among both human and nonhuman organisms. When the Soviet Union tried to break up the nuclear family and to subordinate it to the "*socialist family of nations*," the original benefits of the family were lost. The nuclear family could not have been scaled up to the "socialist family of nations" without losing its essence and the results were quite disastrous, at least for the nuclear family. This case should teach us an important lesson about the limits of scaling a small system and our misunderstanding of scaling (for an interesting discussion see West, 2017). A squirrel that falls to the ground from a high treetop will not hurt itself. Increase its size to that of an elephant and the poor creature will be crushed to death. When analyzing a small system, we should therefore pay close attention to the way identified constraints both organize and empower the system in relation to *specific* tasks and contexts where improvisation is important. A family is a unit of improvisation that can empower its children in a way the socialist family of nations cannot. Therefore, the use of the family metaphor should always be subject to suspicion when it violates the logic of scaling. In such a case, we should always suspect that an act of tyranny is appearing in front of our eyes. The former Soviet Union was definitely not a "family" of nations; the fascist regime of Mussolini, although paying lip service to tradition and family, was not a scaled-up version of an Italian family; and when shallow and manipulative marketing advisors launch a campaign describing a company as a "happy family of workers," this rhetorical move should be exposed, criticized, and rejected. We should also be aware of the point where constraints accommodate to a degree where the system's behavior becomes sub-additive, meaning that "more is less." To better understand constraints, let me discuss a paper that I published with my colleague Dan Vilenchik (Neuman & Vilenchik, 2019).

On Soccer, Mice, and Men

So far in this book, when we have analyzed the behavior of soccer teams, we have analyzed the set of players and the way in which they pass the ball between each other. Each player has been represented through its interactions (i.e. ball passes with the other players). For example, the behavior of a soccer player like Messi (M) may be represented as the normalized vector of his ball passes to the other players in his team. Here are some imagined numbers of ball passes from Messi to Griezmann (G), Alba (A), and Roberto (R) (Table 4.1).

The normalized vector simply means that we convert each number of ball passes into proportions (Table 4.2).

Along the same lines, the behavior of both Messi and Griezmann (M + G) can be represented as the vector of ball passes *both* have made to the rest of the players. Such a procedure of adding together the vectors of the players' behavior is analogous to the graph operation of merging two vertices into a new node whose neighbor set is the union of each vertex's neighbor set.

The complexity of a soccer game cannot be reduced to ball passes. However, as proposed by John von Neumann (see the introduction), science is about modeling, and modeling (as explained in Chap. 1) is inevitably a process that produces a simplified map. Therefore, when we model the complex behavior of a soccer team using ball passes only, we must adopt a modest perspective but nonetheless ask what we can learn from the very simple use of ball passes in our model.

In our paper, we developed a unique methodology for measuring the extent to which the behavior of several players *diverges* from the behavior of the

Table 4.1 The ball-passes vector

Griezmann	Alba	Roberto
70	30	20

Table 4.2 The normalized ball-passes vector

Griezmann	Alba	Roberto
0.58	0.25	0.16

system's components. What does it mean? It means that when we look at the behavior of several players by examining the ball passes they have made to the other players, we may ask to what extent this ball-passing behavior of the players as a whole can be expected based on the ball-passing behavior of the individuals composing the whole. For example, when we examine the ball passes of the composed dyad M and G, we may ask whether we can predict this vector of ball passes by using the vectors of the individual players M and G. Moreover, each combination of two players (such as M and G) can be a subset of a larger combination of players (such as M, G, and A). Therefore, we can also ask to what extent the behavior of each combination of players can be approximated by using the behavior of each superset of players in which it is a part.

Using this methodology, we built a new measure called $_w$, which quantifies the extent to which the behavior of a system of N players diverges from additivity, or contradicts our expectations that are based on the behavior of the composing players. The Hebrew letter $_w$ stands for "shi-tu-fi-yut," which means "collaboration" or "cooperation." Taking each dyad of players, for example, we asked how surprising is its behavior when observed from the perspective of the two *individual* players composing it, and also how surprising it is from the perspective of all three triads of players of which it is a subset. We applied the same procedure to triads of players, tetrads of players, and so on. In each case, we measured the amount of surprise accompanying the behavior of several players when they are observed from the perspectives of their subsets and the higher-level sets in which they are included.

The "granularity" level of our analysis was therefore the number of players composing each level of analysis. For example, we asked how surprising is the behavior of different combinations of two players in a team (e.g. dyads), by examining the individual behavior of the two players composing the dyad. However, we also examined how surprising is the behavior of three-player combinations by examining the behavior of the different combinations of *dyads* composing each triad of players.

We hypothesized that as the *granularity* level of a system increases—for instance, in moving from dyads to triads of players—constraints are accommodated on the potential ball passes and the surprise observed in the behavior of the higher level (e.g. a triad) becomes lower and lower when observed from the perspective of its lower level (e.g. dyads). This surprise was exactly what our new measure represented. The hypothesis that we formed is quite trivial, but when we analyzed the soccer clubs that participated in the 2015–2016 season of the English Premier League, we gained some interesting findings.

The first result was found when we plotted our measure of surprise as a function of the granularity level, or the size of the players' set, that we used. We noticed that, as hypothesized, the surprise evident in the behavior of an *N*-players system decays as the number of players increases. The explanation is simple. When constraints are added, a system becomes more ordered and saturated. When you examine the behavior of a single player, you realize that theoretically he has the ultimate freedom to pass the ball to any of the others. However, when you examine the behavior of several players taken together, then their degrees of freedom decrease because the number of players to whom they may pass the ball is limited and the behavior of a player is constrained by the behavior of the others.

Up to this point there is nothing very interesting to note. The really interesting findings are those concerning the nuances of the constrained behavior. For example, the greatest surprise was evident when we examined the behavior of *triads* of players. The triad (i.e. third-order relations) seems to be a watershed of complexity, an idea that I further develop in this chapter. Moreover, the decay function associating the surprise and the granularity level was one of *exponential decay*or, more accurately, a *stretched-exponential function*. This is an interesting finding as the stretched-exponential function has traditionally been used in physics to describe the returned to equilibrium (i.e. relaxation) of a perturbed system. We hypothesized that this stretched-exponential function is indicative of the way constraints are accommodated and coordinate the behavior of the system as we zoom out globally and gradually from individual players to the whole soccer team. As such, we further hypothesized that the *decay rate* of our surprise measure would be indictive of the extent to which a team of football players is self-organized from the micro to the macro levels of the team. If this hypothesis was grounded, we presumed, then the decay rate of our surprise measure should be correlated with the performance of the team; a team that can better coordinate the behavior of its players across different levels of organization would exhibit a *faster* decay rate of surprise. This is clearly an example of the way in which accommodated constraints are beneficial to a system. And, indeed, when we tried to predict the rank of each team at the end of the season by using the ball-passing behavior of its players and the way it was organized around the team's levels of analysis, we found that the faster the surprise measure decayed, the better was the final rank of the team. The exception was... Leicester City Football Club! Leicester appeared on the graph as a unique *anomaly*. It was ranked at the top of the league, but the behavior of its players was far beyond the ordered behavior of the other groups.

I find this specific result to be interesting as exceptions and anomalies are highly important to the advancement of science. For instance, when examining the spread of COVID-19, we may notice the existence of people who have a Teflon-like character: the virus simply cannot stick to them despite their intensive exposure to sick people who have been infected. This is probably an exception but understanding this exception may have great relevance for medicine and public health, as we may map the genes responsible for this natural immunity and use this knowledge for good. The Leicester anomaly is of the same kind. It seems that Leicester somehow managed to successfully play the game by being less "competent" in quickly coordinating the behavior of single players into dyads, triads, and so on. It is possible that the secret sauce of Leicester was not the ability to be well organized but some kind of unique talent in managing an *ordered form of disorder*, a point to be discussed in the next chapter.

On Love Triangles and the Mysterious Nature of Triads

Another interesting finding of my research with Vilenchik was that we were able to gain the same predictive results regarding the teams' final ranking by just analyzing the five players in each team[4] who played the most—and, even more interestingly, by just looking at the behavior of triads (i.e. third-order interactions) of players among these top five players. That is, to gain the *same* predictive results regarding a team's final ranking, we didn't need to examine the whole team and its different combinations of players, only the behavior of three-player combinations of the top five players. It seems that, like in jazz, the real value of a soccer team can be reduced to trios and their explanatory power.

Interestingly, the number three magically appears in several contexts highly relevant to the understanding of small systems. For example, when analyzing the behavior of mice, researchers have concluded that "the minimal models that rely only on individual traits and pairwise correlations between animals are not enough to capture group behavior, but that models that include *third-order interactions* give a very accurate description of the group" (Shemesh et al., 2013, p. 1, my emphasis).

We may elaborate upon this finding under the title "Of mice and men," echoing the famous novella by John Steinbeck (1937/1993). Mice and men

---

[4] Is it a coincidence that this is the same as the lower bound of Miller's "magical number"—seven plus or minus two (Miller, 1956)?

(or people) are different in many respects, but it is very interesting that in different contexts third-order interactions are key to understanding the complex social behaviors of both. Third-order interactions are interactions in a triad and, surprisingly, we may find these triads almost everywhere. The dramatic triad is a common theme in literature and appears in Shakespeare's works, such as *Hamlet* (Hardison, 1960) but surprisingly in *Romeo and Juliet* too. You are probably familiar with this famous play (Shakespeare, 2004), which describes the tragic love of a young couple. However, you may be unaware of the fact that the romantic dyad of Romeo and Juliet has a third component, which is… Count Paris. You may ask yourself, "Who the fuck is Paris?" who has never registered in our collective memory. The answer is that Count Paris is the person to who Juliet has been promised in marriage. Therefore, what we actually see in *Romeo and Juliet* is a *love triangle* rather than a love dyad per se, and this theme of love triangles appears in numerous places, from *Casablanca* (1942) to *The Bridges of Maddison County* (1995) and *Wuthering Heights* (Brontë, 1847/2020).

Why is the triad a watershed of complexity? West (2017) shows that the total number of pairwise links between N people can be calculated as follows: multiply the number of people (N) by $N - 1$ and divide the output by 2. For two people (for a dyad) the number of connections is 1, calculated as $2 \times (2 - 1): 2 = 1$. For three people the number of pairwise links is calculated as $3 \times (3 - 1): 2 = 3$. Now, when we increase the number of people to 4, we get $4 \times (4 - 1): 2 = 6$ pairwise connections. When moving from 3 to 4 people, we have added just one person but the number of pairwise connections has doubled. There is a nonlinear quadratic function associating the size of a group with the potential number of pairwise connections. When we double the size of the group, the number of these connections approximately increases fourfold. To repeat, if we have a single person then there are 0 connections; if we increase the number by one to two people then we have 1 connection; and if we increase this by another one to three people, we get 3 connections. But, when we move from three to four people, the potential number of connections is doubled. The triad seems to be a watershed of complexity….

The interesting thing is that even complex family dramas that involve more than a triad can be reduced and analyzed through the triads composing them. This is a highly interesting idea—known as *Peirce's reduction thesis*—that was mentioned by Charles Sanders Peirce in a very long paper titled "The Logic of Relatives" (Peirce, 1897).

Peirce first differentiates between first-, second-, and third-order relations. Brotherhood, for instance, is a dyadic second-order relation necessarily involving two people (Peirce, 1897, p. 167). Selling is in contrast a triadic relation, as when we poetically describe an act of treason as "selling their soul to the devil." Peirce hypothesized that higher-order relations, such a fourth-order relations (i.e. tetrads), can be reduced to triadic relations: "That out of triads all polyads can be constructed is made plain" (p. 183). That is, higher-order relations can be reduced to triads with no loss of information. This is not the case with triads. For instance, the proposition "A buys B from C at the price D" (p. 164) can be broken down into two triads: "A buys B from C" and "A buys B at the price of D." Such a process of reducing a proposition to a lower-order relation cannot be applied to triads or dyads. When you talk about brotherhood, you *must* have two brothers—no less. When you say: "I'm sad" or "It's raining," you cannot reduce the meaning of your utterance to a lower-order relation. When you describe the process of selling, you inevitably assume that there is someone who sells, someone who buys, and something that is sold. This relation is inherently much more complex than can be represented by a dyadic relation and cannot be reduced to a dyadic relation.

Peirce suggests that the complexity of higher-order relations can be reduced to triads, but we still have no idea why many small social systems are limited in the number of their particles. Is it because it is extremely difficult to manage the connections and their increased number, as shown by West (2017)? It is still difficult to understand why the triad is such an important level of analysis for understanding small systems and why we find love triangles rather than love tetrads. With the aim of proposing a new explanation—quite embryonic at this stage—let me introduce a famous painting and return to the idea of perspective-taking.

What Can a Famous Spanish Painting Teach Us About Perspectives and Triads?

To understand the deep logic of triads, and to better understand small systems, we now turn to a famous painting—*Las Meninas*—painted by Diego Velázquez in 1656. Here is a partial black-and-white image of this painting (Fig. 4.1).

The painting is fascinating because it involves a play of perspectives. For example, as an observer you see that the painter is located in the left side of the painting. The painter is the one who painted the picture and therefore stands in front of it. However, as the observer who stands in front of the painting, you are paradoxically observing the painter from the same position

Fig. 4.1 *Las Meninas. Source* https://commons.wikimedia.org/wiki/File:Las_Meninas_d etail.jpg

from which he painted the picture. As you are observing Velázquez from the same perspective that he holds in the picture, it seems that both of us hold the same position, which is practically impossible. Moreover, Velázquez is painting the royal couple whose image is reflected in the mirror at the center of the painting. When you understand this then you may suddenly feel a thrill from realizing that the couple that Velázquez is painting, is located in front of the painter, exactly where you stand. When you observe the painting, you observe the painter, who is actually painting himself occupying the same position that you hold and observing something (i.e. the couple) that exists behind (both of) you. This play of perspectives is fascinating but why is it important to understand it? The reason is that perspectives are representations of the world, and through representations we are able to monitor and regulate our behavior both as individuals and as collectives. The painting shows us how the play of perspectives enriches our representation, and from this point we may start speculating about the role of perspectives in small systems. To address this challenge, let me use a linguistic example.

When I say, "I am sad," I communicate my first-person perspective through a linguistic tool, which is the first-person pronoun (singular) "I." Saying that "I" am sad gives others a glance into my own perspective—my inner world. To be able to produce such an utterance, I must be highly reflective. However, we don't have to use the sophisticated linguistic vehicle of the first-person pronoun "I" to communicate our representation. As mentioned before, a bacterium or an immune cell communicating its own perspective is doing a similar thing without using any linguistic vehicle. A perspective is at the most basic level just a representation that is formed through a specific point of view. Continuing with the same linguistic example, we may move from the first- to the second-person perspective. Saying "You look sad" is just a way of providing my second-person perspective on you, or more accurately my perspective on your own representation. It is my representation of your mental state as communicated to you. There is also the third-person perspective, which would be, "He looks sad." This perspective communicates the inner mental state of the person holding the first-person perspective regarding the person holding the second-person perspective, through the mediation of a third person. Therefore, Messi can say, "I am sad"; Alba can say to Messi, "You look sad"; and Griezmann can say to Alba, "Messi (he) looks sad." I'm not familiar with a language with a fourth-person perspective and this fact emphasizes the importance of the triadic relationship.

Previously, I have proposed that the existence of different perspectives, and their integration and coordination by small systems, provides one possible explanation for the unique ability of small systems to improvise. At this point, I would like to further develop this idea and to explain why third-order relations are so important and why we can use our insights to better understand small systems. I will keep using the linguistic example for this purpose.

Let's start with "I." The "I" (i.e. my own first-person position) has a limited perspective on a situation. A soccer player has a limited vision of the field and a member of an unhappy family, such as the mother in *The Glass Menagerie* (Williams, 2011), has a limited and distorted perspective on her family dynamics. We all have a blind spot, which is inevitably formed from the fact that we have a perspective. In order to enrich and complement this representation, the "I" therefore requires a view from the outside. This outside perspective mirrors the way in which I see the world and myself. However, this idea is trickier to understand than you may think at first glance. Let's explain why through the measure of relative entropy.

To monitor and regulate its behavior, "I" requires a good model of its *inner* representation. However, this model might lead to a vicious regression of models where model A requires a model A_1, which requires its own model A_2, and so on ad infinitum. Let's indicate the first-person representation of its inner state by using a probability distribution, P_I. For the purpose of modeling—that is, monitoring and regulating its behavior—"I" requires some *loss function* through which it may calibrate its representation. The loss function is a measure indicating how far the *representation* is from the observed reality, whether the internal reality or the external reality. To better model our inner and outer worlds, we have to compare our model/representation and reality, quantify the degree to which our model diverges from the reality, and calibrate and correct our model through this measure. The relative entropy measure is such a loss function, and we may use it to check our model of ourself. However, by using it from *within*, I will get $D_{KL} = (P_I \| P_I)$, which is a measure that, in a circular way, confirms one's own representation with its limited perspective and blind spot!

The only way of escaping the limited perspective of the particle and its blind spot is by using the outside perspective of others. In contrast with some misleading individualistic beliefs, there is no "I" without others (Neuman, 2020). The other's representation of "I" may be expressed by the probability distribution Q_Y. In this case, communicating the outside perspective with "I" may take the form of the relative entropy score of $D_{KL} = (P_I \| Q_Y)$, where my representation of your mental state can be modeled as a distribution of mental states used to approximate your true distribution of mental states. The only possible way in which "I" may represent, regulate, and correct its own representation is by *modeling itself from the outside*, through the other. This process of "identity formation" may take the form of the following composite function:

$$1_A = D_{KL}(P_I \| Q_Y) \circ D_{KL}(Q_Y \| P_I)$$

or, simply stated, this means, "I can see myself only through the way in which you see me." Whenever there is a gap and

$$1_I \neq D_{KL}(P_I \| Q_Y) \circ D_{KL}(Q_Y \| P_I)$$

then I know that my representation should be adjusted. The problem is that gaps in perspectives are inevitable. The relative entropy measure, being asymmetric, wonderfully represents this aspect, because the only case in which

$$D_{KL}(P_I \| Q_Y) = D_{KL}(Q_Y \| P_I)$$

is when "I" and "You" have exactly the same representation of me, but as each of us holds a *different perspective*, by simply occupying a different position in the world, such a situation is impossible. The asymmetry of perspectives is almost by definition built into the definition of a perspective and the asymmetry of the relative entropy measure perfectly represents it.

The asymmetry of perspectives cannot be resolved through "mutual gaze" as each of us holds a different perspective with its own blind spot. Here the third-person perspective comes into the picture. The third communicates to the second what it sees in the first (e.g. "He is sad"). The three different perspectives formed by a triad of particles are therefore like the three *non-collinear points* required to define a plane. Non-collinear points are points that are not located on the same line. To define a mathematical plane, three such points are required. Analogically speaking, the *minimum* of *three non-collinear perspectives may be required to define the plane of meaning* that is essential to each of the points/particles for monitoring and regulating their behavior. Let me explain why.

When communicating its own representation to the second person, "I" may gain a convergence of three different points of view: its own representation, its representation as communicated by the second, and its representation by the third as communicated to the second. Through this process of *triangulation,* "I" may model and regulate itself, and attune (i.e. adjust) its representation to those of the others. However, the real explanation as to why third-order interactions are so important may come from the fact that the relative entropy measure that we used to model the asymmetry of perspective-taking has a unique property: it does not satisfy *triangle inequality!*

Triangle inequality means that for any triangle, the sum of the lengths of any two sides must be greater than or equal to the length of the remaining side.[5] For the relative entropy measure, this means that the triangle inequality:

$$D_{KL}(P||Q) \leq D_{KL}(P||U) + D_{KL}(U||Q)$$

is not satisfied.

Let's illustrate this idea through the use of three perspectives: first (F), second (S), and third (T). Have a look at the following Fig. 4.2, where arrows indicate the direction of the relative entropy measure. For instance, the directed arrow from T to F signifies $D_{KL}(F || T)$, which means that a third person is approximating the distribution of F's mental states.

In the left-hand diagram, we see that T "factors through" S. Triangle inequality means that when T approximates F directly (e.g. "He is sad"),

---

[5] https://mathworld.wolfram.com/TriangleInequality.html.

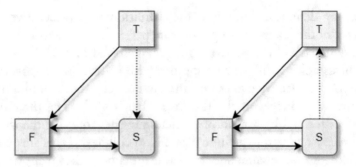

Fig. 4.2 Perspective-taking diagrams

it will probably get a different result from the one it gains when approximating F's representation *through the perspective* of S. This means that S isn't *redundant* to T and that the perspective of T cannot be reduced to that of S. Similarly, the right-hand figure shows that T isn't redundant to S. In sum, F, S, and T have limited perspectives that can be covered through their mutual perspectives. First, F cannot see itself from within and needs the perspective of S. S has a limited perspective on F, as evident from its different positioning (e.g. I am not You) and expressed by the asymmetry of the relative entropy. Using the perspective of T, S may cover up its own blind spot and the same holds for T, which doesn't hold any reciprocal relations with F. When adjusting its own representation, F may therefore use three different coordinates:

1. $| D_{KL} (S \| F) - D_{KL} (F \| S) |$ (the gap formed between F's representation and the one formed through S—the other)
2. $| D_{KL} (F \| S) - [D_{KL} (F \| T) + (T \| S)] |$ (the gap between the way S sees F and the way S sees F through the eyes of T)
3. $| D_{KL} (F \| T) - [D_{KL} (F \| S) + (S \| T)] |$ (the gap between the way T sees F and the way T sees F through the eyes of S).

These three "coordinates" are the minimum required to calibrate the representation of each particle, and any other higher-order interaction may be reduced to a triadic structure as proposed by Peirce. We may conclude by saying that triads form the minimal perspectival constraints through which a particle may monitor, regulate, and adjust its behavior with relation to others. This explanation is actually a hypothesis or speculation that may explain why third-order relations are so important in modeling small systems, from families to soccer teams, and why when we scale up these systems, they lose their unique character. Bringing too many perspectives into the system may drain the unique perspective of the individual particle by subordinating it to the perspective of a single particle located at the top of the hierarchy, such as the conductor of a symphonic orchestra, or by averaging the individual perspective through the "wisdom" of the mob. Its voice may be lost in the forest of perspectives.

Summary

- Constraints are "boundary conditions" imposed on the behavioral space of a small social system.
- Perspectives may serve as natural constraints that guide the behavior of the system.
- A perspective is a representation that is formed through a specific point of view.
- Triads form the minimal perspectival constraints through which a particle may monitor, regulate, and adjust its behavior in relation to others.
- Three non-collinear perspectives may be required to define the plane of meaning through which a small system is formed.

5

How Messy Should You Be to Win a Soccer Match? On the Yin and Yang of Entropy

In this chapter, we understand why order is so important but also why injecting enough disorder is crucial for a small social system. We also learn how the contribution of the defenders in soccer is unjustifiably underestimated, why entropy is closely associated with flexibility and the performance of a small social system, and how to explain pictures to a dead hare.

On the Fantasy of Perfect Order

There is a wealth of scientific evidence suggesting that we are inclined to see order even in places where it does not really exist. Think about the shape of clouds. How many times have you watched a puffy cloud and its mass of droplets take the shape of a familiar figure, such as a sheep or a face? The cloud as a mass of droplets has a clear physical existence with its unique forms of organization. However, when we see a face reflected in a cloud, it is our mind that imposes order on this mass and for a good reason: order is what allows us to survive and in the context of small social systems it is what allows us to collaborate and work together. In this context, the annihilation of disorder seems to be a sacred mission as it is disorder that threatens the heart of our existence. However, in a world governed by irreversibility, fluctuations, competition, and cooperation, the fantasy of *perfect* order is not only impossible but dangerous. Striving for perfect order is accompanied by

© The Author(s), under exclusive license to Springer Nature Switzerland AG 2021
Y. Neuman, *How Small Social Systems Work*, The Frontiers Collection,
https://doi.org/10.1007/978-3-030-82238-5_5

an unbearable price and soccer is a wonderful context for illustrating this point. Let's explain this important idea by focusing on ball passes.

In the previous chapter, I explained the concept of entropy through ball passes. Let's imagine a matrix of ball passes between the players of a team. Each player is generally aware of the matrix, at least as it concerns his own limited perspective. In the case of maximum entropy, the ball passes are evenly distributed. This means that the chances of each player passing the ball to any other player are the same. This is an imagined situation expressing disorder and it has detrimental consequences for the ability of the players to function as a team. When each player is facing such uncertainty, he cannot plan his steps and coordinate with the others—for instance, by signaling that he is free of his opposing player and available to receive the ball. In contrast, a situation where the distribution of ball passes deviates from homogeneity maximizes the communication and cooperation between the team's players. For example, knowing that the probability of a pass from Messi to Suárez, who are both forward players, is significantly higher than the probability of a pass from Messi to Ter Stegen, who is a goalkeeper, may strongly support Barcelona's coordinated behavior when Messi is holding the ball.

We now understand that a team has a clear and basic interest in minimizing the entropy of ball passes in order to maximize the communication between its players. However, in an adaptive environment, order has its Achilles' heel. The first problem is that an ordered pattern of ball passes may expose a team to counterattacks by the opposing team. By knowing your patterns, your opponent may use them against you. For every move there may be a countermove, and for each strategy there will be a counter-strategy. In other words, the first problem in forming an ordered behavior is that despite its crucial importance for constituting efficient communication *within* a team, it is a point of vulnerability *between* teams, because an opponent may use the ordered pattern of its "prey" for his own use.

For example, knowing that there is more likely to be a ball pass from Vidal to Messi than from Vidal to Griezmann, the opposing team (e.g. Sevilla) may relocate its resources to more powerfully block Messi when Vidal has the ball. This is of course a dynamic situation that must be managed in real time. A team playing against Barcelona can allocate all its resources in order to block Messi from getting the ball. However, in this case, other talented forward players, such as Suárez, might use their relative freedom to score. Therefore, the allocation of resources during a soccer match is a dynamic process that responds to the real-time situation (e.g. Messi is holding the ball) by taking into account known patterns of behavior (e.g. he is probably going to pass the ball to Suárez).

Minimizing the disorder *within* the team comes with the potential price of exposing the team's vulnerability to the other team, and vice versa. Therefore, when examining small social systems, we should pay close attention to the way in which teams manage the delicate balance between order and disorder. This conclusion is not only valid for soccer but also for the behavior of many other systems. Think about the patrols performed by the Israel Defense Forces (IDF) along the country's northern border with Lebanon, which is an area governed on the Lebanese side by Hezbollah, considered even by the EU to be a terrorist organization. The inherent interest of the IDF is to maximize the simplicity of launching patrols in order to save energy and communication efforts. For example, the patrols may follow a well-known path along the border. This ordered behavior clearly makes it easier for the soldiers to manage the patrols by turning them into a habit. However, this routine may make them vulnerable to attacks by the enemy, who may observe and learn these patterns. For example, having become familiar with the patterns of the IDF, Hezbollah launched an attack on an Israeli patrol in 2006, killing several soldiers and kidnapping others, an incident that led to the 2006 Lebanon War.

The attack on the Israeli patrol illustrates the in-built danger of being too ordered when operating in a competitive and adaptive environment. Order has a clear advantage when individuals are trying to communicate with each other, but it has a clear danger in a competitive environment, whether a football match, a military conflict, or a poker tournament. However, it is not clear whether this danger holds for other—generally non-competitive—small systems that we have discussed, such as families or jazz trios. To explain the relevance for systems such as families or jazz trios of being messy enough, I would like to introduce another main difficulty that is associated with a too-ordered behavior.

There is an inherent pitfall in maximizing the order within the system, which is the pitfall of *limiting the system's degrees of freedom and its ability to flexibly adjust, adapt, and bootstrap itself to avoid dangerous states of stagnation.* This point can be explained through a concrete example and through models formed in physics.

Let's imagine a boutique advertisement company where a team of creatives (i.e. people whose job it is to suggest new ideas) is engaged in a new campaign for cat food. To launch the process, the team starts by trying to form creative associations with "cat." Associated words such as "kitten," "dog," and "rabbit" immediately pop into the minds of the creatives, as in our mental lexicon or semantic network—i.e. the network of words stored in our mind—they are located near to "cat." Now, if our semantic memory can be modeled as a network composed of words and their meanings, how can we model the way in which we move from one word to another? The distance between two words can be a function of their use in the same context. For example, in the USA, "hotdog," which is a kind of sausage, will be close to words such as "ketchup," "mustard," and "bun." If you are a Berliner and your favorite sausage is the *Currywurst*, then this kind of sausage will be associated in your mind with "curry," whereas an American may have a significant association between "curry" and Indian cuisine (rather than with a sausage).

In a semantic network, moving from one word to the next may be a function of the distance between the words, which in turn may be a function of their co-occurrence (i.e. their neighboring appearance in the same lexical context). However, this simple dynamic of traveling along the paths of a semantic network may limit our ability to identify interesting and creative associations.

Think about the word "French." In the USA, it is mostly associated with the word "fries"[1] as both words appear in the compound "French fries." Additionally, it turns out that the relationship between the words "French" and "fries" is symmetric, and the word "fries" is statistically most associated with "French." If we stick with the law according to which we move from one word to another based only on strength of association, then whenever we hear the words "French" or "fries," we will find ourselves *trapped in a cycle* between those two words. This cycle is called an *attractor* as it describes a limited region toward which a dynamic system evolves. In our example, the dynamic of associative thinking is such that when we start from the word "French" we move to the word "fries" and are trapped forever in a cycle limited to two words of the whole semantic network. This idea can easily be extended to any other movements (or "walks") through a network, whether it is a network of words, ideas, or physically located sites.

We can think about such a walk through a semantic network as a *deterministic walk* and compare it with a *random walk*, where our movement from one word to the next is determined by chance only. Risau-Gusman, Martinez, and Kinouchi (2003) studied a stochastic walk where the "walker" was allowed to move to its neighbors in the network with varying transition probabilities. To model such a process, they used a parameter of *temperature* (i.e. T) that determined the stochasticity level of the walk. When T is very small (approaching zero), the traveler always goes to the nearest neighboring site, eventually trapped in a cycle with a pair of neighbors (Risau-Gusman

---

[1] https://www.english-corpora.org/coca.

et al., 2003). This is exactly the case described above where the dynamic of associations is trapped in a vicious cycle composed only of "French" and "fries."

Rigid thinking or behavior, whether on the individual or the group level of analysis, can be described as a dynamic that is limited to certain *attractors* and trapped in vicious cycles without being able to achieve release from them. As a young student of psychology, I once volunteered in a hostel for mentally disabled individuals. As I was a former art student, I was invited by the social worker managing the place to deliver an art class to the hostel's residents. I decided to use a very simple exercise that I had enjoyed as an elementary school boy. The exercise was supposedly very simple. I sketched a simple geometric figure, such as a circle, on the board and asked the residents of the hostel to copy the figure and develop it into a painting using their *associations* only. For example, when I sketched a circle, the associations that immediately came into my mind were a basketball, a plate, a pitta bread and so on. So, for instance, given the figure of a circle, a resident could use it to paint a basketball. The idea was simple but after a couple of minutes, I noticed that the residents, who were grown adults, were baffled. Each of them stared at his or her board holding a pencil in his hand, unable to move forward. I approached one of them and asked him to look at the circle and to imagine what it looked like. The resident was helpless. Nothing popped into his mind. To encourage him, I asked whether the circle looked to him like a pitta bread, which is commonly encountered in Israel on a daily basis. Although more than 30 years have passed since my experience with this group, I still remember this disabled person's surprise and excitement when it was revealed that the circle could remind him of a pitta bread. By himself, he could not have thought of this similarity. For him, a circle was just a circle. No more, no less. In retrospect, I believe that such a process can be modeled as the most extreme case of a deterministic walk, where there is no walk at all from one mental "site" to another. These mentally disabled individuals were trapped in their own limited, simple, concrete, and highly rigid attractors.

When modeling movement in a network, Risau-Gusman et al. (2003) found that when the temperature parameter T is higher than 0, some interesting things start to happen. The temperature or the stochasticity level may allow us to escape the traps formed by the constraints, expressed in the example of the semantic network by the strength of associations (i.e. the semantic distance) between words. The team of creatives working in the advertisement company may find out that deterministic walks cannot take us beyond the known and that random walks, which are indifferent to real-world constraints, produce irrelevant garbage associations. To generate a real creative process, whether in art, music, or soccer, a team must increase the "temperature" of the system to allow the optimal level of entropy, noise, and disorder. This may release the rigid structure and has the potential of moving the small system of creatives to a new place. Although disorder has a bad reputation, in the right amounts and at the right times, it is necessary for the existence of any small system.

What is true for a football team is also true for a family. Think, for example, about a family working within highly rigid patterns of behavior. Such a family appears in *American Beauty* (1999), featuring Kevin Spacey and Annette Bening. Beyond the various sociological and psychological interpretations of the movie, what we see are families caught in a rigid dynamic where the roles imposed on the individuals leave them imprisoned in misery. This dynamic is epitomized by the character of Colonel Frank Fitts, who imposes strict discipline and order on his family to an extent that not only causes his family extreme misery but also finally leads to a tragedy.

Highly ordered systems or states are less flexible in regulating their behavior and responding to real-time challenges. Like crystals, they express impressive order and potential stability but clearly lack the resilience and anti-fragility (Taleb, 2012) that are so important for living systems. In a case where the system is big enough, the resilience may be the result of an evolutionary process where those who did not meet the high bar of a changing reality simply did not survive. However, such a process is a luxury available only for very big systems, such as the population of a virus. This luxury is not available to a small system such as a football team or a family operating on a small scale and for a short period of time.

In the context of a small system, a careful balance must be maintained between order and disorder. Yin and yang must be balanced not in the simple symmetric and static sense, as visually represented by the yin-yang symbol, but by carefully adjusting order/disorder in real time in order to produce tailor-made solutions. A paper I co-authored with my colleagues (Neuman et al., 2018) aimed to test the importance of order/disorder in soccer by focusing on ball passes, and I would like to present our main findings in order to further discuss the importance of the order/disorder balance in small systems.

Measuring the Disorder of a Soccer Team

In our study, we used a dataset describing the behavior of the teams participating in the 2015–2016 season of the English Premier League, the season where Leicester surprised almost everyone (see Chap. 2). For each of the participating teams, we computed the entropy of ball passes using a measure called *Tsallis entropy*.[2] Here is the measure:

$$Sq(pi) = \frac{1}{q-1}(1 - \sum pi^q)$$

You can see that the measure has two components: the probability of the event (denoted as p_i) and some kind of *parameter* (denoted as q). The *parameter* or *index* may take whatever values are chosen by the researcher. This parameter can be set up to "amplify" rare events in the distribution and to overweight their contribution to the general entropy score. It can also be set to overweight more common events in the distribution.

For example, let's assume that we are measuring the entropy of individuals' choice of ice-cream flavors in order to predict a personality trait such as openness to experience. Our hypothesis is that those who are ready to experience less conventional flavors are people who also score higher on a personality dimension known as Openness. The most popular ice-cream flavors, at least in the USA, are chocolate (which is the favorite of 17% of people) followed by vanilla (15%) and strawberry (8%).[3] Assuming that most people, whether they are open to experience or not, favor conventional flavors such as chocolate, we need to consider whether to give extra weight in the final entropy score to the more or the less conventional favors. For example, let's take a hypothetical distribution where 80% of the ice cream consumed by a person is vanilla and the rest is an obscure flavor, such as meatballs (20%). When measuring the Tsallis entropy of this distribution with an index of $q = 0.2$, we get 0.85, and when we measure the entropy with an index of $q = 2.0$, we get 0.32. An entropy index of 0.2 amplifies rare events in the distribution and gives them more weight in the final entropy score. If we are interested in predicting openness to experience based on the distribution of ice-cream flavors, and if we know that there are a very few dominant flavors (e.g. chocolate), then we may want to give more weight to the rare flavors.

In our soccer study, we used Tsallis entropy to emphasize the relative contributions of rare and common events in our analysis of ball passes. For example, in our first analysis, we used an entropy index of 0.2, as this index amplifies "rare" events, which in our case includes ball passes where forward players are involved. As ball passes involving forward players are relatively

---

[2] https://en.wikipedia.org/wiki/Tsallis_entropy#:~:text=In%20physics%2C%20the%20Tsallis%20entr opy,the%20standard%20Boltzmann%E2%80%93Gibbs%20entropy.

[3] https://today.yougov.com/topics/food/articles-reports/2020/07/14/popular-ice-cream-flavor-poll-sur vey-direct.

rare, amplifying such rare events gives them extra weight in the final entropy score. This may have consequences for identifying patterns in the behavior of a soccer team that may be correlated with the team's success, as will be shown in the following paragraphs.

We first tested the hypothesis that the rank of a team at the end of a season (i.e. its success) can be predicted using a single simple measure, which is the team's entropy score for its ball passes. This hypothesis does not require too much explanation. An ordered pattern of ball passes should be indicative of a team's ability to efficiently communicate and also monitor and regulate its behavior. To be efficient, order is a must.

We measured the entropy of each team's ball passes by using two parameters of the Tsallis entropy index ($q = 0.2$ and $q = 2.0$) and with Shannon's entropy ($q = 1.0$), as these measures represent (1) the common Shannon entropy measure; (2) a *super-additive* entropy index ($q < 1.0$), which amplifies the probability of *rarer* events; and (3) a *sub-additive* entropy index ($q > 1.0$), which amplifies the probability of more *common* events.

We found that the higher the entropy of a team (the more disordered it is), the lower is its position at the end of the season. The lower the entropy of the ball passes, the better the team's rank at the end of the season. This finding is easily explained because a higher entropy score means that the pattern of the team's ball passes is less organized, and a less organized team is a less successful team.

If you recall the opening chapter and the reasons why Leicester was not considered a likely candidate to win the 2015–2016 Premier League, then you will realize that one of the most basic predictors of a team's final rank may be its position at the end of the previous season. At this point, you may ask yourself whether the entropy measure contributes to the prediction of a team's final rank above and beyond what may be gained through knowing the team's rank at the end of the previous season. When we added the entropy measure to our predictive model that included the position of the team at the end of the previous season, we found that our prediction improved by roughly 17%, which is quite a lot. Therefore, a team's level of orderliness seems to add to the prediction of its success far beyond what can be gained from its recent history of success (or otherwise).

Order in Art and Soccer

The association between a team's level of orderliness and its success is not surprising, but it may refute some naïve ideological biases that dismiss the importance of order. When we analyzed the data, I recalled my days as an art student within the changing zeitgeist between old and new schools of art. My old teachers emphasized the importance of technique: being a painter was primarily about being able to paint. In contrast, the newer and younger school of teachers were dismissive of the order imposed on matter through a painter's skill and mastery. Art was about personal expression, and technique was secondary at best. For example, the level of technique shown by Marcel Duchamp in memorable pieces such as *Fountain* was minimal. After all, what technique is required to present a urinal in a museum? Or take another example. One of my young teachers' heroes was an artist by the name of Joseph Beuys (1921–1986). In one of his well-known performance pieces—*How to Explain Pictures to a Dead Hare* (*Wie Man dem Toten Hasen die Bilder Erklärt*), which was his first exhibition in a private gallery—Beuys let the gallery visitors watch him explaining pictures to a dead hare. In another of his pieces—*Green Violin* (*Grüne Geige*)[4]—Beuys simply displayed a green violin. In both of these cases, no technical mastery was required, and the order imposed on matter through the artist's skill and talent were irrelevant. Art was not about order but about rebelling against the order formed by the older generation.

The same approach could lead us to naïvely expect football to be the art of performance, surprise and personal expression. Whether FC Barcelona maintains an ordered pattern of ball passes could be conceived as less important than the "performance art" and artistic and personal expression of talents such as Lionel Messi. However, it seems that order, whose specific expression is context dependent, is a necessary aspect of human conduct, whether in art or sport. Freedom of expression cannot be detached from the order formed through the technical mastery that allows and enables the freedom of artistic expression. The order imposed through language allows poets to express their creativity just as the order of ball passes allows a football team to express its creativity. It is the delicate balance between order and disorder that matters, and not the dominance of one over the other.

The Hidden Heroes of Soccer

So far, the major finding my colleagues and I describe in our paper is not too impressive unless you arrive at it with a dismissive perspective on the role of order. However, interestingly, we found that the best prediction of a

---

[4] https://www.moma.org/collection/works/100550?sov_referrer=artist&artist_id=0&page=2.

team's success using the entropy measure for the ball passes was gained using the entropy index of 0.2. This parameter amplified the probability of *rarer* ball passes. This means that when we privileged rare events in the ball pass matrix, which was precisely the point of the super-additive index, our prediction of the team's success significantly improved. How can we explain this finding and why privileging rare ball passes in the general entropy score of a team improves the prediction of its final rank at the end of the season? This finding may hint at the role of certain types of player—such as forwards, who are involved in fewer ball passes than other team members—in influencing a team's position at the end of the season. Let me explain and test this hypothesis.

In soccer, most ball passes are conducted in the midfield by midfield players. The most common events in the distribution of ball passes are passes between midfield players, who are responsible for moving the ball toward the opponent's goal. Ball passes involving defenders and forwards are less common. Therefore, my colleagues and I hypothesized that the rare events privileged by the super-additive Tsallis index are associated with the defenders and/or forwards, and more specifically with their ordered behavior relating to ball passes. Moreover, we hypothesized that if a team's performance, as measured through Tsallis entropy, is best represented through the more ordered behavior of the *forwards*, then it should be expressed in the ability of the Tsallis entropy to predict the *goals scored by the group*. After all, the forwards are "formally" in charge of scoring goals. In contrast, if the team's success more heavily leans on the *defenders' ordered behavior*, then we should see the effect of the team's performance while trying to predict the number of goals *conceded by the team*.

We tested these hypotheses by measuring the correlation between the entropy measure and either goals scored or goals conceded by each team. We found a positive and statistically significant correlation between the entropy measure and goals *conceded*. No correlation was found with goals *scored*.

I find this result to be of great interest as it may point to the important and overlooked role of the defenders in coordinating the ball's movement and contributing to the team's success. It is reasonable to hypothesize that defenders who are more coordinated in moving the ball forward are more coordinated in defending their goalkeeper from an approaching attack. The finding is of interest because the glory and prestige in soccer are almost exclusively given to the forwards.

As of the time of writing, Nélson Semedo, a defender at FC Barcelona, earned a yearly salary of £3,796,000, while Luis Suárez, a forward, earned £27,664,000. Although differences in experience, talent and reputation may explain this salary gap, it seems to validly represent the conceived contributions of forwards versus midfielders and defenders. While goals scored by forwards can easily be measured and these players' relative contributions to the success of a team can easily be identified and understood by the layperson, it seems that we fail to realize and appreciate the importance of defenders and their significant contribution, as a *coordinated small system* within a team, to the success of the whole team.

As I explained in the first chapter, small social systems are different from populations and individuals, and the success of a team cannot simply be reduced to the virtue of individual players. The success of a small system is dependent on the synergy of its composing particles, and one important consequence for readers of reading this book may be a reassessment of the relative contributions of the individuals composing such a system. Again, while the praise in soccer is usually given to those *individuals* who score goals, the analysis by my colleagues and I may point to an interesting finding, which is the important contribution of the defenders' collective and their "negative" success in avoiding goals.

We usually underestimate the contribution of "negative success." We praise those who lead our country to victory in a war while ignoring those who prevent a war. We admire the Navy SEALs who killed Osama Bin Laden—the arch terrorist—while ignoring the work of those who prevent terrorist attacks. And, similarly, we glorify those *individuals* who score goals while barely noticing the hard labor of the *collective* of players who prevent the players of the other team from scoring goals.

The Yin and Yang of a Small System

So far, we have studied the orderliness of a football team. However, we have considered the orderliness within the team while ignoring the adaptive aspect of the team's behavior as the team confronts its opponent. To address this aspect, my colleagues and I measured the entropy of each team in every game and compared it to the entropy of the opposing team in the *same* game. We calculated the difference in entropy between the competing teams in each match, and for each team we computed the sum of the entropy differences across the season. After computing the average entropy difference of each team, we computed the ratio between this score and the entropy measure of the team's ball passes matrix. We hypothesized that the new score, expressing the degree to which a team presents itself as less ordered than it is, would be correlated with the team's success. Indeed, we found that the higher the "adaptive entropy" score (i.e. the entropy of the team *in comparison* with the entropy of its opponent), the better it was positioned at the end of the season. When we built a new predictive model incorporating the team's rank in the previous season, its entropy score and its adaptive entropy score, we found that the *only* significant predictor of the team's success at the end of the season was the adaptive entropy score.

The general lesson that we may learn from the analysis is that a more organized team is a more successful team. However, this finding should not be confused with rigidity. This is because the less organized the behavior of a team in a specific game (as compared with its opponent), the less expected its behavior is by the opponent, the higher are the team's degrees of freedom, and the higher are the chances of the team both scoring and avoiding goals. I emphasize this finding as we usually think of organization and disorganization as opposites. However, here the competency of a soccer team may be expressed (paradoxically, one might say) in forming an ordered pattern of ball passes while within this order exhibiting a relatively higher level of disorder that expresses the team's relatively higher degrees of freedom in moving the ball.

From the Tardigrade to FC Barcelona

In this chapter, I have pointed not only to the importance of order but also to the importance of disorder in maintaining a small social system. While my main focus has been on the behavior of soccer teams, the same general logic seems to pervade all living system of which we know. A surprising example can be found in the behavior of the *tardigrade*. The tardigrade, also called the water bear, is a micro-animal known for its incredible ability to survive in hostile conditions, including dehydration, radiation levels 1000 times higher than what other animal can stand, and extreme temperatures. It was found that a tardigrade-unique protein by the name of Dsup (damage suppressor) protects tardigrade DNA from damage during exposure. No less interesting is the more recent finding that "the protein is intrinsically disordered," which enables it to "adjust its structure to fit" the DNA's shape (Mínguez-Toral et al., 2020).[5] Being "intrinsically disordered" may have clear benefits in the case of a potentially traumatic exposure to damage, and in the case of

---

[5] I am talking here about conformational entropy: https://en.wikipedia.org/wiki/Conformational_entropy.

the tardigrade it allows the protein to have maximal flexibility in adjusting and protecting the DNA. This is exactly the same process that I've discussed before, involving maximizing entropy under constraints and increasing the requisite variety of a system.

Having a fluid-like structure implies higher flexibility and, again, we find this idea echoing throughout many fields and phenomena. The highly intelligent octopus may escape from a shark by fluidly sneaking through small apertures and an open-minded individual can fluidly adjusts his cognition to avoid dogmatic thinking and functional fixedness. In fact, it is argued that brain entropy is associated with divergent thinking and creativity (Shi et al., 2020), a finding that supports the importance of entropy in the flexible adjustment of various systems. A team trying to survive in an unexpected and extreme environment, such as the team led by Shackleton during his Antarctic expedition (Shackleton, 1919/1999) or a soccer team facing a challenging opponent, is better off losing its rigid schemes so it can fluidly adjust to real-time challenges. Small systems are potentially well suited to handle such situations, as a small system may "naturally" increase its entropy as a result of constrained communication. The "noise" inevitably formed through the constrained interactions may have clear benefits as long as the price of anarchy is balanced by the benefit of order. The behavior of a large population of particles can take the form of erratic movements that, when averaged, cancel out each other's "bias" to potentially create a Gaussian distribution and convergence toward the mean, which is not always the optimal situation. It can also result in the particles moving into a rigid and simple structure where they are aligned, as is evident in herding behavior. In a small system, such as a soccer team, "errors" cannot cancel each other out (on average), and the contributions of individual particles and their topological information content as a whole are evident—for good or bad.

In this context, we may think of several potential directions to take to diagnose and "heal" a dysfunctional system, such as a dysfunctional family or soccer team. The first direction is to measure the system's adaptive level of order/disorder in a changing environment. For example, we could measure the extent to which and the rate at which the system becomes more disordered in stressful situations. We could also measure the degree in which the system is able to increase its entropy so as to flexibly adjust to stress and bootstrap itself from harmful attractors. Based on the abovementioned finding regarding the underestimated contribution of defenders in football, we can also ask to what extent, defenders function in an orderly manner when they are under attack and who are the weakest links in the chain of defense. We can further ask whether the team tends to stagnate under stress, and which are the "anarchistic" players who, when they enter the game, may inject a sufficient amount of chaos to release the team from its trap or attractor. Such individuals may be substitute players who most of the time sit on the bench but have the talent of injecting the right amount of *constructive noise* into the game. Agents of chaos do not necessarily have a negative influence on the behavior of a small system, and in the right context they may make a crucial contribution to the adaptive behavior of a team. Being open-minded seems to be an imperative of life—an ongoing challenge to small systems struggling to maintain the right balance between order and chaos.

Summary

- In a world governed by irreversibility, fluctuations, competition, and cooperation, the fantasy of perfect order is not only impossible but dangerous.
- Although disorder has a bad reputation, in the right amount and at the right times it is necessary for the existence of any small system.
- Entropy is highly important in the flexible adjustment of various systems.
- The "noise" inevitably formed through constrained interactions may have clear benefits as long as the price of anarchy is balanced out by the benefits of order.
- Agents of chaos do not necessarily have a negative influence on the behavior of small systems, and in the right context they may make a crucial contribution to the adaptive behavior of a team.

6

Unique Interactions: On Chemistry in Love and Soccer

In this chapter, we discuss chemistry in love and soccer through the romantic comedy *As Good as It Gets*. We also learn how to use the idea of a mathematician and Nobel Laureate in order to (1) measure the relative contributions of players to a game, (2) identify sneaky academics, and (3) measure a team's synergy. We also learn, through the movie *Good Will Hunting*, how chemistry can influence the individual and how Jaynes' method of hypothesis-testing may be used to model a small social system.

As Good as It Is

Listening to sport commentators, or psychology "commentators," discussing small social systems such as soccer teams or couples, one cannot avoid the conclusion that we have a strong essentialist bias that leads us to believe that human beings or soccer teams can be reduced to *inherent* individual traits or properties that fully explain their behavior. It is as if a successful soccer team has a kind of merit built into it and a person's behavior can be explained by a singular in-built personality trait. In the context of personality, the psychologist Mischel (2004) challenged this dogma many years ago by pointing to the *contextual* aspect of human personality.

The more you delve into the study of psychology and small social systems, the better you understand the poverty of the essentialist approach and how people like Walter Mischel and Gregory Bateson were right in pointing in a

© The Author(s), under exclusive license to Springer Nature Switzerland AG 2021
Y. Neuman, *How Small Social Systems Work*, The Frontiers Collection,
https://doi.org/10.1007/978-3-030-82238-5_6

different direction. In the study of small systems, a more constructive method is to look at the interactions between the particles and the way in which they unfold in *time* and respond to *contexts*. The secret sauce of a small system exists "in between" the particles and can be interpreted only when we take context and dynamics into account. This is a very challenging and demanding approach but, in this chapter, I would like to use it to discuss the meaning of "chemistry" between people and its importance in understanding the behavior of a small system.

Chemistry might be described in terms of good *rapport*, which is defined as a close and harmonious relationship of mutual understanding between people. The etymology of "rapport" directs us to the taxonomic categories of relationship, information and communication.[1] Having good chemistry with someone implies the ability to harmoniously communicate with minimal effort. The Principle of Least Efforts seems to stand at the heart of chemistry. In contrast, bad chemistry or bad rapport means a deep sense of misunderstanding and a lack of basic communication. Chemistry, if it exists, can therefore come in two distinct flavors. However, as we learned in the previous chapter, ultimate chemistry, despite its harmonious nature, might be a poisonous ingredient in human relations. This harmony may lead to a degree of smoothness that leads people to work mechanically, so that boredom gains the upper hand in relationships or rigid disorder imposes itself on the system.

In contrast with human beings, molecules have no chemistry, at least in the sense of the complex emotional and psychological interactions between two people. They swirl in huge clusters of gas or liquid, showing no evidence of preferable or unique interactions with "significant others." While to the best of our knowledge chemistry is not observed between gas particles, this doesn't imply that chemistry necessarily appears among human beings. In fact, it is seldom observed among human beings. Even within our very limited social circle, very rarely do we describe our relationships with others in terms of chemistry. In some cases, our relationships are more similar to those of the gas particles. Chemistry is rare and unique. It exists only in small social systems, and it may be one of the important ingredients that makes a small system effective in flexibly responding to real-time challenges.

As a group-level characteristic, chemistry is impossible in a large group of people, as implied by the impossibility of maintaining reciprocal connections between many individuals (see Chap. 4). Chemistry therefore exists only for

---

[1] https://www.oed.com/browsethesaurus?thesaurusTerm=rapport&searchType=words&type=thesaurus search.

interacting individuals, for small groups of individuals or for certain "coalitions" of individuals working within a social system. My argument is that chemistry cannot be scaled up and its potential existence in a small system may explain some of the unique characteristics and potential benefits of a small social system. We will start by discussing and modeling the important aspects of having unique interactions and "chemistry" through the romantic comedy *As Good as It Gets* (1997), featuring Jack Nicholson and Helen Hunt.

As Good as It Gets is a charming romantic comedy. The film describes a successful writer by the name of Melvin Udall, portrayed by Nicholson, who has a unique talent for playing troubled figures. Although he has a successful career, Melvin is a total failure in his personal life and interpersonal relationships. He suffers from a severe form of obsessive–compulsive disorder (OCD) and is also clearly a lonely, bitter and unhappy person. His loneliness stems not only from his disturbing OCD but also from his misanthropic behavior and the fact that he simply cannot maintain healthy and positive interactions with others. The exception that pops up in the film is a single mother and waitress by the name of Carol Connelly, who serves him his daily breakfast at the local restaurant. Carol is a patient and compassionate person, and through her complex relationship with Melvin, she enables him to feel love, become happy and restore his creativity. However, it is a long road and it starts quite negatively....

The first scene where Carol and Melvin interact is in the restaurant, where Melvin is shocked find that "his" table is occupied by two other diners. In the context of the previous chapter's discussion of order, we understand that disorder is the most threatening experience facing a person with OCD. Melvin approaches Carol and without any greeting announces that he is "starving" and complains that there are "Jews" at his table. Carol's response is firm. "It's not your table," she says. "Behave." Although Carol reprimands Melvin for his rude and racist behavior and threatens that he will not be allowed to visit the restaurant again, she also allows herself to comment on his bad diet from a *caring* perspective. Melvin's response to her caring approach is not just cynical but vicious, as it refers to her sick child, who is her soft spot. Carol is emotionally shocked by Melvin's response and she aggressively puts him in his place, threatens him and insults him by saying that he is a "crazy freak." Thus far, this short interaction doesn't look like the opening of a promising romantic relationship but it also doesn't seem like typical dialogue between a server and a customer.

Short functional interactions in such places where breakfast is served (e.g. a restaurant) usually have a well-defined *script*. The server opens with greetings (e.g. "Good morning, how are you today?") followed by a question (e.g.

"What would you like for breakfast?"). The customer is supposed to respond to the greetings politely and to concisely answer the question by placing his or her order. Carol seems to be less formal in her approach to her customers. For example, we see her warmly greeting a female customer and calling the customer's little daughter an "angel." In contrast with Carol's warm approach, Melvin is an antagonistic person, who clearly deviates from the script required of the polite customer. Both Carol and Melvin deviate from the script but in diametrically opposing ways. However, despite the differences between Carol and Melvin, even in their first encounter there seems to be a hint that theirs is a *unique* interaction; it starts from bad chemistry but evolves toward a positive romantic relationship by the end of the film. How is it possible to model and measure the uniqueness of Carol and Melvin's interactions?

We may use a simple procedure. First, we can represent Carol's interactions with different customers by using a topical analysis tool such as LIWC2015 (Pennebaker et al., 2015). This tool classifies the words in a text into predefined content categories. For example, the words used by Carol in her interaction with the mother-customer can be reduced to several content categories, such as Positive Emotion, Social, and Family. In contrast, in his first interactions in the restaurant, Melvin makes references to Body Parts (e.g. the noses of the "Jews" sitting at his favorite table), Death, and Food as he deliberately emphasizes the high level of cholesterol of his order (which, among other things, contains sausages) as a kind of comment challenging and teasing the accepted social norms emphasizing the importance of healthy food.

Carol's interaction with Melvin is different from her interaction with the mother-customer, as she expresses anger and uses derogatory language (i.e. "crazy freak"). We can represent Carol's interaction with the mother-customer using a vector (i.e. an array of numbers) describing the distribution of her topical content words. The same procedure can be applied to any other customer using with whom Carol interacts.

For simplicity, let's assume that in her interactions with customers, Carol uses words that fall under the following content categories: Positive Emotion, Negative Emotion, Swear Words, and Family. We average and normalize her use of content categories in her interactions with customers (Table 6.1).

Table 6.1 The distribution of Carol's topical words in her interactions with customers

Positive emotion	Negative emotion	Swear words	Family
0.8	0.02	0.01	0.17

We can see that Carol is usually a very positive person in her interactions with customers. Most of her language use falls under the category of Positive Emotion. At this point, we should measure the distance between Carol's interaction with Melvin and Carol's averaged interactions with other customers. We can then normalize the distance by dividing it by some kind of disparity measure, such as standard deviation.

Let's form the normalized vector of topics Carol is using when interacting with Melvin (V1: C → M) and along the same line the vector of topics Melvin uses when interacting with Carol (V2: M → C), the topics characterizing Carol's interactions with other customers (V3: C → O) and the vector of topics characterizing Melvin's interactions with other waitresses (V4: M → O). We may now use some kind of a distance measure D between the vectors or the distributions. The uniqueness measure of the interaction between Carol the waitress and Melvin the customer can be simply defined as follows:

$$\text{Uniqueness}\,(M\_C) = \frac{1}{2}(D(V1, V3) * D(V2, V4))$$

This simple measure can give us an idea of how unique the interaction between Carol and Melvin is. It scores the extent to which Carol's interaction with Melvin is different from her interactions with other customers and the extent to which Melvin's interaction with Carol is different from his interactions with other waitresses. There is no point in comparing apples to oranges, so the relevant benchmarks for measuring the uniqueness of Melvin and Carol's interactions are Carol's interactions with other *customers* and Melvin's interactions with other *waitresses*.

As chemistry between individuals is relatively rare, it seems that chemistry implies uniqueness. Therefore, chemistry seems to have an aspect that is beyond smooth communication of the kind that requires minimal effort. The habitual script of ordering in a restaurant expresses a ritualized form of communication that is highly efficient. However, no one mistakes their efficient interaction with a waitress as "chemistry." Saying that I have a chemistry with a specific waitress just because our communication is smooth and efficient doesn't grasp the deep meaning of chemistry among human beings. Chemistry is exclusive, like intimacy, and as such it implies uniqueness.

That said, chemistry cannot simply be reduced to uniqueness. At first, the communication between Melvin and Carol is unique but doesn't seem to involve chemistry at all. Only later, when romantic relations develop, can we say something about chemistry. It therefore seems that chemistry has two specific aspects. The first is uniqueness and the second is synergy, here used in the sense of a surprising and positive outcome emerging out of interactions and forming a new unit of analysis; the coupe Carol and Melvin for instance. In sum, uniqueness means that an interaction has a signature that marks it as different from other interactions serving as benchmarks for comparison. The synergy aspect means that something new, a whole different from the sum of its parts, emerges through this unique interaction. In this context, it seems interesting to ask what is so unique about uniqueness.

"Unique" is a descriptive term. We can measure how unique is a given interaction using the above procedure. The meaning of uniqueness, though, should explain to us what is unique about being unique. The first answer is that a unique interaction increases the information content of the interacting couple and forms a boundary around them, marking them as "different." As the interaction between Melvin and Carol is unique, it marks their dyadic relationship as different, and this boundary seems to be a necessary condition for the construction of a "couple": Melvin–Carol. Their existence as differentiated structural unit—a couple—can be illustrated by the informative value the outside observer may attach to this couple, saying, for instance, "I could never have believed that a misanthropic person such as Melvin would fall in love" or "I could never have believed that such a nice person as Carol would fall in love with a freak like Melvin." Beyond intuition, we may better model the chemistry of dyads or triads; the next section presents a novel methodology for doing so.

On the Shapley Value, a Gluttonous Cat, and Sneaky Academics

Chemistry is evident not only in romantic relationships but also in a good piece of scientific work. In my experience at least, there are cases where you find yourself retrospectively reflecting on the process that led to a good academic publication, saying to yourself, "I could not have done it without my partner." If your partner is reflective enough, he or she will probably think the same. As you can see, chemistry is a clear characteristic of small social systems that maintain significant interactions, whether in love, academia, or soccer. However, in the academic context there are cases where the contributions of the various authors to the publication of an academic paper are clearly not the synergistic product of collaboration and chemistry. There are in fact cases, known to almost any academic, where sneaky co-authors did not really contribute to the paper. In some cases, you may doubt whether they even read the paper before becoming a co-author. How is it possible to model and measure the synergistic chemistry of the "players" in cases such as these? In order to propose a possible methodology, I first outline the idea of the *Shapley value*.

The Shapley value was introduced in the context of game theory by the Nobel Laureate Lloyd Shapley. It allows us to measure the contribution of a player to a game and the payoff that he or she should reasonably expect given his or her contribution.[2] Let's introduce this idea by using two examples. I would like to start with a very simple example that involves three cats.

At a certain point, my family and I had three domestic cats adopted by my young children: a mother cat, whom my daughters called Kitty, and its two male offspring, whose nicknames were Jake and Cat. Each of the cats had a different physique and consumed a different amount of food. However, as an observer by nature, I noticed that when they ate together, different coalitions of the cats consumed different amounts of food. When the mother cat dined with one or two of her offspring, she consumed much less than when she ate alone. However, when the two male cats ate together, they consumed much more than when they ate alone or with their mother. Observing these interactions, I recalled the idea of *social facilitation* and the way in which the presence of others may influence even the consumption of food.

Now, let's assume that the three cats dine together from the same plate and on average consume 15 kg of cat food over a given period. On average and when eating alone over the same period, Kitty (God blessed her soul) consumes 1 kg, Jake 2 kg, and Cat 4 kg. When we try to understand the "shares" of each cat when they dine collectively, it is clear that their relative contributions are different. First, we cannot say that each cat consumes one-third of the food, as we know that they each consume different amounts of food and also that they consume different amounts of food when eating in different coalitions of cats. We may measure the different amounts of food consumed by the different subgroups of cats:

As shown in Table 6.2, the *marginal contribution* of each cat to a coalition (or subgroup) is then calculated as the output of the subgroup minus the output of the same subgroup excluding the individual participant. For example, the marginal contribution of Kitty to the dining coalition of Kitty + Jake is 3 − 2 = 1. The Shapley value of Kitty is therefore calculated as her *average marginal contribution* across all of the differently sized dining coalitions. Her contribution as an individual player is 1, her average marginal

---

[2] A nice illustrative example appears at https://www.bis.org/publ/qtrpdf/r_qt0909y.htm#:~:text=The%20Shapley%20value%20of%20each,5%20(see%20bottom%20row).

Table 6.2 The Shapley values for the cats' relative contributions

Subgroup	Subgroup output	Marginal contribution of Kitty	Marginal contribution of Jake	Marginal contribution of Cat
Kitty	1	1	–	–
Jake	2	–	2	–
Cat	4	–	–	4
Kitty + Jake	3	1	2	–
Kitty + Cat	5	1	–	4
Jake + Cat	14	–	10	12
Kitty + Jake + Cat	15	1	10	12
Shapley value		1	6	8

contribution to coalitions of two cats is 1, and here contribution to the three-cats coalition is 1 too. On average her Shapley value is 1. The Shapley values of Jake and Cat are 6 and 8 (rounded) respectively.

Through the use of the Shapley value, we see that when the cats dine together, Kitty is responsible for 7% of the food consumed while Jake is responsible for 40% and Cat for 53%. We can also see that when they dine together, the family of cats consumes significantly more than they consume when each of the cats dines alone. This is a result that may be attributed to social facilitation, but it is also important to notice that the relative contributions of each cat may teach us something interesting. When Cat dines alone, he consumes 4 kg, but when he eats in a coalition of cats, he consumes twice as much (i.e. 8 kg). In comparison, when Jake dines in company, he consumes three times more than he consumes when eating alone.

Jake is a thin and sportive cat with a strong competitive urge, whereas Cat is fat and lazy. By calculating the Shapley value, we may learn that although Cat is known in our family as a glutton, the scaling of his appetite to the dining coalitions indicates that he is much less influenced by social facilitation than Jake, the thin and competitive cat, who generally consumes much less when eating alone. This conclusion teaches us something interesting about the family of cats as a small social system. However, the example was intended simply to introduce the idea of the Shapley value. I now turn to another example taken from a different field: the academic field, as mentioned at the opening of this section.

This second example illustrates the Shapley value through the "academic game" of publishing. It assumes the existence of N players-academics collaborating on a paper published in an imaginary and prestigious journal in the social sciences which is called *Scientific Interpretations*. We start by assuming that the paper has three co-authors: YN, OP, and SH. OP is a young and ambitious academic striving to get a university position. When the paper is published, he immediately applies to an academic position, proudly mentioning his paper and presenting it as an indication of his personal virtue.

The committee discussing his candidacy is impressed. After all, in the academic culture of the social sciences and the humanities where quality is not trivial to measure, what can be better than relying on the quality mark of the journal in which the paper has been published? This is the same logic through which people are judged on the basis of their alma matter—the university from which they graduated—rather than on their own unique achievements. In fact, I have personally encountered this behavior several times and for some reason always by graduates of Harvard. It has always surprised me that the educated people of such a fine institute would express the mentality of the old world, where people were judged not as individuals but as members of social "clans." For example, imagine yourself going back in time and meeting the noble King Charles II of Spain. "I'm a noble member of the Habsburg family," he may have declared with pride. Luckily, he would not also have been able to declare himself a graduate of Harvard… However, being a "graduate" of the Habsburgs would not have been a good quality mark for judging the quality of Charles the individual, as poor Charles, whose parents were family relatives, suffered from several severe defects, both mental and physical. Chap. 3 previously explained why the Nazis were wrong in their understanding of entropy, and the poor king in this story is just another instance of the argument previously presented. Now, let's return to our committee and their enthusiasm regarding the *Scientific Interpretations* paper.

The enthusiasm of the committee is questioned by a critical member. This member says, "It is an impressive achievement, no doubt. Publishing in *Scientific Interpretations* is a quality mark like being a Harvard graduate, which I must modestly admit that I am. However, how do we know the *relative* contribution of OP to the paper?" This is a perfectly legitimate question and one deserving a serious scientific answer. Not all members of the committee understand the issue (after all, most of them are not Harvard graduates …). The critical member explains: "It is possible that the achievement of OP may be largely attributed to the talent and efforts of his co-authors and that his own contribution to the success was marginal if any." The members of the committee are perplexed. It is the first time a university committee has been faced with such a problem—the justified demand to check the relative contribution of an individual in a rational and scientific way. As professional academics, they may even have forgotten that the *raison d'être* of their existence is the scientific approach and as such they should adhere to this approach. At this point, the question is whether it is possible to measure the relative contribution of OP.

One possible direction for answering the above question is to measure the contributions of OP to each subset of authors who contributed to the science paper. In our case, we have three authors (OP, YN, and SH) and the following combinations of dyads: OP + YN, OP + SH, and YN + SH.

Next, for each "player" and for each subset of "players," we assign a value (v) that represents the success of the player(s) in the specified game. In the academic case, we can measure the value of each player through the *impact factor* of the journals in which each of the researchers/dyads publishes. Impact factor is simply a scientometric measure that describes the importance of a journal. Like almost any other measure known in the social sciences, it has its own difficulties and shortcomings; however, for our illustrative purpose, these difficulties are not of much importance. What is important is that we have a single measure of a journal's importance and we may measure the contribution of OP to the success of publishing a paper in an important journal. To do this, we may measure the average impact factor of the journals in which each of our academics (YN, OP, and SH) has published as a solo author, and the average impact factor of the journals in which each combination of our authors has published.

We can then try to determine the contribution of OP to the two subsets of which he is a part (OP + YN and OP + SH) and also average OP's marginal contributions to the "coalitions" of academic teams of which he is a part. Let's assume that the success of the academic "players" and their coalitions is as follows (Table 6.3).

As YN, OP, and SH have published together only once, their success value is actually the impact factor of the journal *Scientific Interpretations*, which has score of 42. However, we can see that when SH publishes alone, the journals

Table 6.3 The values for the sneaky academics example

S	v(S) (i.e. impact factor)
SH	1
OP	1
YN	12
YN + SH	3
OP + SH	1
YN + OP	32
YN + OP + SH	42

in which he publishes are ranked low, having an average impact factor of 1. The same holds for OP, while the situation for YN is much better.

To repeat what has already been explained above, Shapley proposed a formula that may be used to measure the relative contributions of each player to a game. Using it in our example, we may try to calculate the relative "shares" each of the above academics should get of the 42 points gained from the *Scientific Interpretations* publication. This situation is totally analogous to the one you have probably observed in action movies where a bunch of thieves have just robbed a bank and are arguing over the most appropriate way to distribute the loot. In action films, this "academic" dispute usually ends with someone pulling a gun rather than by anyone loading up a formula containing the Shapley value. It could be quite amusing to observe a bunch of tough guys pulling their guns, when suddenly one of them says, "Hey guys—instead of fighting, why not use the Shapley value?" I guess that this wise guy would be the first to be shot....

In the academic example, we can use the formula proposed by Shapley. When we apply this formula to the above data and round the results, it turns out that of the 42 points of the paper, 23 points should go to YN (55%), 17 points should go OP (40%), and only 2 points should go to SH (5%). Having evaluated the relative contribution of OP to the paper, the university committee may feel that while OP has made a contribution to the paper and deserves his own share, he can by no means prove that his contribution was such that he deserves all of the credit or even the majority of it.

This is a useful way of measuring the relative contributions of individuals, but it is important to be aware of potential difficulties in implementing the above procedure. For example, in the real world, not all coalitions actually exist, and therefore we cannot measure their values. For example, YN and SH may never have published a paper together and therefore the coalition YN + SH wouldn't exist. In addition, computing the Shapley value for a large number of players may lead to computational overload. These difficulties can be addressed in the study of small social systems, but for smoothness of presentation let's assume that we may proceed using the Shapley value in a straightforward way and by remembering that it is mainly used to illustrate how we can measure the contribution of an individual to some kind of a small system of players.

From Shapley to Synergy

In the previous section, I introduced a common measure to determine the relative contribution of each "player" and his or her fair share in the outcome of a collective activity (e.g. publishing an academic paper). However, when studying small social systems, our interest is not in measuring the relative contributions of each player but in measuring the *synergy* of several particles. So, how can measuring the relative contribution of a single player to a coalition help us to measure the synergy of several individuals?

One possible way to address the challenge of measuring synergy is to move from the mathematical realm of game theory to the softer realm of machine learning (ML). ML has been a hot topic in the past few years, and it is commonly discussed in the context of artificial intelligence. Putting the buzz aside, we should understand what ML really is. There are various definitions of ML, but I believe that the best way to think about it is in terms of algorithms and methodologies for prediction that work through the automatic optimization and self-improvement of a function. This may seem too general, but a simple example may easily clarify what ML is all about. Let us return to the family of cats: Kitty, Jake, and Cat.

Let's assume that a small door, the size of a cat, allows the three cats to get in and out of the house. However, the door can be used by other cats to enter the house—a situation that we want to avoid. An automatic device opens the door whenever one of the three cats arrives. This device could use chips inserted into each cat's collar. However, we are tech geeks and would like the door to open when a camera authenticates a cat through facial recognition. In this context, the solution must include some kind of algorithm that, when fed with the visual image of a cat's face, can successfully identify whether the cat belongs to the family or whether it is another creature, such as an intruding cat or a raccoon.

ML can offer us the following procedure. First, we collect a very large dataset of pictures. Some of the pictures are of our domestic cats and the others are of potential intruders. Next, we represent each picture as an array of pixels. This is the input fed into our algorithm. The next phase is to divide the dataset into a learning set and a test set. In the training phase, the algorithm uses a given mathematical model that includes some variables, or as they are called in ML "features," and their values. The algorithm learns to assign weights to the model's features in order to achieve the best prediction accuracy possible.

For example, when teaching my introductory course in cognitive sciences, I could try to build a ML model to predict who is going to fail or pass the exam. As features, I might use variables such as effort invested in the course, active participation, hours of sleep before the exam, and so on. During the training phase, I would present the algorithm with the measurements collected for every student on the course and ask it to tune the weights of the model in such a way as to gain the best predictive results.

In the example of the cats, the features are the pixels and their different combinations. During the test phase, I present the algorithm with pictures taken by the small door's camera and ask it to perform a binary classification of the pictures into (1) our domestic cats and (2) others, based on the model that it learned during the learning phase. The core of the ML approach is *optimization*, which is a concept already discussed in Chap. 4. In the context of ML, optimization concerns the way in which some kind of a loss function is minimized through the delicate tuning of the model's parameters.

Now let's see how ML can be used to measure synergy. We will start with a simple example—the challenge of determining the synergy of two academics we have met before: YN and OP. First, we identify a large enough dataset of academic authors in the academic field in which YN and OP publish. In our dataset, we have all cases in which papers have been published by two authors and the impact factors of the journals in which they were published. During the learning phase, we train a linear regression model to predict the impact factor of a journal in which a given dyad published. This is our dependent variable—the target or the criterion. To predict the dependent variable, we use the marginal contributions of the individual authors in a procedure analogous to the one presented in the above examples illustrating the calculation of the Shapley value. In other words, we use the Shapley value of each author as an independent variable and use it to predict the impact factor of the journal.

During the test phase, we examine how good is our model at predicting the joint activity of each new dyad of authors it didn't see in the learning phase. If the performance of the model reaches the required level of performance, then we may be satisfied and move on. Now we get into our specific dyad of authors: YN and OP, a couple our ML algorithm did not meet in the learning or test phases. Knowing that OP's marginal contribution is X and YN's marginal contribution is Y, we may use the linear model learned during the learning phase to predict the outcome of their joint efforts in publishing a paper.

A simple linear model has the following form:

$$Y = a + b * X$$

where X is the independent variable, a is a constant indicating the intersection of the regression line with the Y axis, and b is the slope of the regression line. Using a large corpus of data to predict the joint activity of dyads based on their individual contribution, we may try to predict the joint impact factor of OP and YN based on their individual contribution. This simple example can be extended to more than two "players," but thus far we haven't touched upon the synergy aspect. Now, in a case where our prediction *deviates* from the actual score of the dyad OP + YN, there may be two major sources of this deviation. The first source is error. No one is perfect, to include our ML models, and even if our model is generally successful, errors in prediction are inevitable. The second source is synergy. To recall, we are using an *additive* model, which proposes that the joint outcome of the two players can be predicted through the simple sum of their individual values. However, if a deviation exists from an additive model then it may be the result of synergy, which is a kind of interaction that leads to an outcome that cannot be predicted by an additive model successful as it may be. Following this logic, we may measure the synergy of the players (in this case, OP and YN) by using the following line of reasoning:

IF YN and OP have no synergy/chemistry
THEN $Y(YN + OP) = a + b1 * YN + b2 * OP$
However, because $Y(YN+OP) \neq a + b1 * YN + b2 * OP$
AND
The error $E = (Predicted\ (YN + OP) - Actual\ (YN + OP))\ ^2$ is significantly higher than the size of error to be expected in predicting N players' coalitions ($Ɇ$) of a similar size
THEN the joint activity of YN + OP presents a synergy to a degree (E/Ɇ) * 100

The synergy of N players can therefore be expressed through the above syllogism, which should not be mistaken for a recipe but as only one instance of one possible approach to modeling the synergy of N "players."

So far, we have presented one possible way to measure the synergy involved in writing an academic paper, but what about soccer? Or dyadic interactions such as those between Melvin and Carol? The only obstacle facing the implementation of the above line of reasoning is defining the value (v) of an interaction. In the case of Melvin and Carol, we can simply measure the uniqueness of their interaction using the procedure previously described. In the case of soccer, the situation may be a little bit more complicated. We can represent the behavior of each dyad of players using the relevant interaction scores, such as the numbers of successful and unsuccessful ball passes between them, the path lengths of the ball passes between them leading to a goal, and so on. By measuring the behaviors of each relevant dyad, we may try to build a model to predict the number of goals scored (and conceded) in a match. Some interactions between players may be found to be of no relevance to the predictive model and others may be found to be relevant to different degrees. The respective weights of each dyad in the model predicting success are the outcomes of our predictive model, and these weights are used as inputs for the second phase. During the second phase, the weight associated with each dyad in the model and indicating its contribution to the success of the team, is used as a dependent variable, which we try to predict given a set of relevant features of each player composing a dyad or simply the Shapley value of each player. In this case, too, a deviation from the prediction may signal synergy, whether positive or negative.

This is one possible (and novel) approach to measuring the synergy of individuals at different levels of granularity in a small system. Measuring synergy means measuring the chemistry that exists in a small system, but there is more than one way of thinking about chemistry. There is more about this in the next section.

On Chemistry and Change

As I have previously argued, a significant interaction cannot be reduced to a unique and smooth communication. Chemistry is about the synergy produced by a small social system of interacting individuals. However, chemistry may also be expressed in interactions where a change is evident in the behavior of the interacting components. For example, in *As Good as It Gets*, the interaction between Melvin and Carol is life-changing for them both. It is these life-changing interactions that stand at the heart of romantic comedies, from *Pretty Woman* (1990) to *Notting Hill* (1999) and *As Good as It Gets* (1997).

Another example is the drama film *Good Will Hunting* (1997), featuring Matt Damon as Will Hunting and Robin Williams as his psychologist, Dr. Sean Maguire. Will is a young working-class guy from South Boston who does a series of low-status jobs, such as construction and cleaning. However, Will is also a math genius, although his talent is unrecognized even by him. By chance, this talent is exposed to a math professor at the

Massachusetts Institute of Technology (MIT)—Professor Gerald Lambeau—who is a winner of the prestigious Fields Medal. Prof. Lambeau is overwhelmed by Will's brilliance and takes him under his wing in order to promote his talent. As Will has a background of angry outbursts and problems with the law, Prof. Lambeau asks for the help of his old friend, the therapist Dr. Maguire, and Will starts participating in therapeutic sessions under Maguire's guidance. Chemistry may not be what springs to mind when we observe the interactions between Will and Dr. Maguire. Will challenges his therapist, argues with him, and even brutally hurts his feelings—an injury that Dr. Maguire responds to in a way one would not expect from a therapist. Like the drama and tension evident in *As Good as It Gets*, Will and Dr. Maguire's interactions are not characterized by smooth and efficient communication. The interactions that they experience are unique but not smooth and efficient. However, as we would expect from a good drama, there is a turning point, which is Dr. Maguire's response to Will's vicious and personal attack. At this point something changes. Will start trusting Dr. Maguire, exposes his traumatic childhood with a violent and abusive father, and start experiencing a process of healing and self-compassion while Dr. Maguire also experiences a turning point in his own life.

Chemistry in small systems, from dyads to soccer teams, doesn't necessarily require smooth, pleasant, and efficient communication. The interactions between Melvin and Carol and those between Will and Dr. Maguire are *troubled* interactions, at least at the beginning. However, they should not be judged by their sweet flavor (or lack thereof) but by their unique signature and the way they change the lives of the participants. Through his interactions with Carol, Melvin learns to love and to accept his deficiencies. Although his OCD doesn't change, his way of living with his severe disorder does. At the end of the film, there is a scene where Melvin notices that he stepped on a crack in the pavement, something that he obsessively tried to avoid in the past. However, instead of reacting to this accident with horror, he simply accepts it and enters a shop with Carol. This is a radical change that can only be traced to the healing interactions between the two. Similarly, Will, who experiences a significant interaction with Dr. Maguire, is subsequently able to struggle with his own demons and to take the courageous step of leaving his childhood neighborhood and working toward a new, promising future. Chemistry is therefore not only about unique and synergistic interactions but also about how such interactions unfold over time in a way that brings about a qualitative change in some or all of their participants.

When we study the chemistry of several players, we should ask ourselves not only whether their interaction is unique and whether it is a whole different from the sum of its parts, but also whether this interaction has changed the way the players behave, as individuals within a team. Two or three soccer players have "chemistry" if their interactions may lead to the formation of new opportunities for each player and to game-changing behavior. Scoring a goal as a result of an interaction may be evidence that the players involved in the interaction have some kind of chemistry. The evidence for such an instance of symmetry should not be intuitively accepted but scientifically validated by showing that a well-defined and unique pattern of interaction leads to a different behavior with a higher probability than the probability measured when observing this behavior given a different kind of interaction. This sounds a little complicated but let me explain by modeling interaction through Jaynes' approach to hypothesis-testing and assessing the weight of evidence.

Let's assume that pre-theoretically—that is, before any theoretical consideration—we conceive Messi and Suárez as a dyad of players with great chemistry. To test the hypothesis that they have great chemistry, we can ask what is the probability of a goal being scored (our hypothesis, H) if Messi and Suárez maintain a minimal sequence of ball passes (i.e. the evidence, E) in a well-defined time-window (e.g. two minutes) before a goal is scored. This is a form of Bayesian reasoning where we measure the probability (P) of a goal being scored *given* the preceding interactions between the two players [i.e. $P(H/E)$]. Notice that this probability may be a simple and good measure of synergy, much simpler than the one described in the previous section. If the probability of scoring a goal, given the unique interaction between two players, is high, then we may conclude that these players have chemistry. However, to analyze the chemistry of the players, we should also ask what is the probability of a goal *not* being scored given the *same unique interaction* [i.e. $P(-H/E)$]. To analyze the degree of success of the interaction between Messi and Suárez, for example, we may measure the posterior odds of scoring a goal *given* a certain pattern of interaction between the given players (i.e. E):

$$\text{Posterior Odds} = \frac{P(H/E)}{P(-H/E)}$$

In order to determine the degree of success of the interaction between Messi and Suárez, we may measure the probability of a goal being scored given their preceding interaction, compared with the probability of a goal *not* being scored (i.e. $-H$) given the same form of preceding interaction between Messi and Suárez. We may then examine these posterior odds in light of the posterior odds of other coalitions of players.

A way of measuring posterior odds was proposed by Jaynes (1996/2003)—the same Jaynes who was mentioned before in the context of maximum entropy. An amusing exposition of this measure appears in a blog post written by Kurt (2015) and the measure is explained and used in several of my papers (e.g. Neuman et al., 2019, 2020).

Jaynes' method of hypothesis-testing can be used to test the hypothesis that a dyad or a triad of players has chemistry. My friend Saadia Gozlan, who is a football fan, argues that wonderful chemistry was evident between Karim Benzema, Luka Modrić, and Cristiano Ronaldo when they all played together at Real Madrid football club. There is no difficulty in testing his intuition, or any other intuition regarding a triad of players, by simply applying the above procedure. If chemistry exists in a small system of players, then it should be expressed in some behavioral outcome (e.g. scoring a goal) that follows a unique interaction between the players. By computing the probability of a goal being scored given a unique interaction characterizing a set of players and by comparing it to a relevant benchmark, we may measure the synergy level of a small system of players. The chemistry of a small system may therefore be expressed through the extent to which an individual may benefit (i.e. score a goal) through the unique interactions between the players. If chemistry was really a signature of Benzema, Modrić, and Ronaldo's interaction then our analysis should reveal a tendency for them to maintain interactions (e.g. ball passes) that lead to a positive outcome (e.g. scoring a goal).

"As Time Goes By" is a famous song played in the memorable film *Casablanca* (1942). This song reminds Rick Blaine (played by Humphrey Bogart) of his love for Ilsa Lund (played by Ingrid Bergman), but also reminds him of her alleged betrayal. Therefore, the song is forbidden in his café. The song does not only bring up a painful memory. When Rick and Ilsa unexpectedly meet in Casablanca, the song becomes a promise: as time goes by, things may change, with the possibility of Rick and Ilsa becoming lovers again. Time is not only associated with memories of the past but also with the possibility of change. As time goes by, new opportunities arise through interactions with others. Some of these interactions are unique and are characterized by chemistry, meaning that the interactions result in a new behavior of the composite whole and/or a change in the behavior of some of the individuals composing it or those associated with it. In this chapter, I have tried to clarify the meaning of chemistry and to propose several ways of modeling chemistry. Being able to measure the chemistry of individuals composing a small system may help us to better understand where and how to interfere in order to improve the behavior of the system. Imagine that a soccer team is performing poorly but there is no clear explanation of why. The source of failure may be attributed to the team, but wouldn't it be much more effective to try to map the specific interactions that fail? Or the contexts where certain interactions fail? Think about Will and Dr. Maguire. The life-changing event that Will experiences cannot be solely attributed to the talent of the psychologist or to the talent of Will and his remarkable intelligence. There is a single situation or context where a phase transition is evident in their interaction. What happens in this specific interaction and why does it lead both Dr. Maguire and Will into a phase transition? Learning from such "diagnostic situations" may lead us to a better understanding of soccer teams, dyads, and families, as long as we remind ourselves to move on from indispensable intuition to a no less dispensable quantitative and empirically based process of modeling.

Summary

- Chemistry is usually defined in terms of harmonious interactions between individuals.
- Chemistry has two specific aspects: uniqueness and synergy.
- Chemistry is also expressed in interactions where a change is evident in the behavior of the components of an interaction or those influenced by it.
- This chapter introduces a simple measure of uniqueness.
- The Shapley value allows us to measure the contribution of players to a game and the payoff that each player should reasonably expect given his or her contribution.
- This chapter presents a couple of ways to measure the synergy of several individuals working as a small system.

7

The Weakest Link in the Chain: From the Colossal Failure of Brazil's Football Team to Sir Earnest Shackleton's *Endurance*

In this chapter, we are introduced to the famous football match between Brazil and Germany at FIFA 2014, and we try to understand, through the Anna Karenina principle, what is the unique signature of happy families. We also use Liebig's law of the minimum to attempt to explain why some football teams fail, learn about Sir Ernest Shackleton's heroic expedition to Antarctica in 1914, and understand how a team of 28 explorers survived against all the odds. Through this case study, we learn about the importance of flexibility and propose a new way to identify the weakest link in the chain and its impact on soccer teams in particular and small social systems in general.

"Cracking Up"

The 2014 FIFA World Cup took place in Brazil, hosting 32 soccer teams all competing for the title. The host—Brazil—was a favorite to win. Given the team's history of success in the World Cup, its well-known football heroes of the past (such as Pelé and Sócrates), its reputation of playing beautiful football, and its talents (such as Dani Alves, Thiago Silva, and Neymar), the odds in favor of the "samba team" winning the title were high. The betting odds marked it as the favorite to win, with Lionel Messi's Argentina as the second favorite and with Germany heading the European favorites, but with its odds to win lower than those of Brazil. Brazil beat teams including Chile and Columbia on its way to the semi-final, where it met the German team. The

© The Author(s), under exclusive license to Springer Nature
Switzerland AG 2021
Y. Neuman, *How Small Social Systems Work*, The Frontiers Collection,
https://doi.org/10.1007/978-3-030-82238-5_7

match[1] took place on July 8, 2014. Although a favorite, Brazil was missing some of its key players, whereas Germany had its own talented players, such as Toni Kroos, Thomas Müller, and Mesut Özil. However, the odds of each team winning the match were considered to be relatively equal.

The game opened with a Brazilian blitz, which threatened the German goal twice within the first three minutes. The first impression was that the Brazilians had clear dominance and that the odds favoring them winning the competition were clearly justified. However, 11 min into the game, Müller scored the first goal for Germany. It was a clear failure of the Brazilian defensive players, who left Müller free during a corner kick. The Brazilians continued to play very well, forming the impression that the goal they had conceded was a local rather than a systemic failure. However, 23 min from the beginning of the game, Miroslav Klose scored the second goal for Germany. A Brazilian female football fan was seen crying in the audience. Later, Kroos scored the third goal for Germany, and another shocked Brazilian football fan was seen in the audience with her mouth open in surprise. Two minutes later, Kroos scored the fourth goal and this time, a little Brazilian boy was seen crying in the audience. Brazil 4, Germany 0. Then, 29 min from the beginning of the game, Sami Khedira scored the fifth goal for Germany. During the second half of the game, André Schürrle scored two additional goals for Germany. The Brazilians managed a single goal in the 90th minute. The game ended as Germany 7, Brazil 1.

It may not be surprising that the German football team won the match, but it is quite surprising that the game ended with the result of 7–1 to Germany. This six-goal difference was a statistical anomaly as it was the worst loss of a host country in the history of the World Cup and the biggest ever winning margin in a semi-final or final. For Brazil this was not just a loss but a painful defeat. To understand how painful this defeat was for Brazil, we must understand that in Brazil, football is a religion no less than a sport. Putting pain aside, it seems that in retrospect everybody realized that something really bad had happened to the Brazilian team. The coach of the German team— Joachim "Jogi" Löw—said that his players were extremely cool, realized that the Brazilians were "cracking up," and took advantage of this breakdown (BBC, 2014).

Beyond the fascinating drama of the match, we may ask ourselves an interesting question, through which we may learn an important lesson for life. The question is why small social systems break down under pressure, and a complementary question is how other small social systems may overcome a

---

[1] https://www.youtube.com/watch?v=jW5jobEpkk4&list=WL&index=13.

crisis against all the odds. These questions and the lessons we may learn from them are the focus of this chapter.

Anna Karenina, Liebig's Law of the Minimum, and the Danger of Cracking Up

Brazil's loss to the German football team should not have been a surprise. In soccer, the game may end in one of three possible results: win, lose, or draw. In certain advanced phases of several tournaments, a draw is not allowed; in the case of a draw, extra time is added or a penalty shootout is used to decide who is the winner. In this context of a binary result, it is always the case that one of the competing teams wins and the other loses. In such cases, if we are fully ignorant and know nothing about the competing teams, we may assign them equal probabilities of winning or losing the game. As there must be one of only two possible outcomes to the game, it should not be a shock to learn that one of the teams lost and the other won. As devoted fans of the losing team, we may be highly disappointed by the result, cry in agony, or complain about the bad luck of our team, but, from a rational perspective, we cannot a priori deny the possibility of a loss or its reasonable probability. However, losing is one thing and defeat is another. When Brazil lost to Germany, it was not the loss that was so surprising but the six-goal difference. This unlikely result was probably what led the German coach to describe the behavior of the Brazilian teams in terms of a breakdown.

An improbable and emotionally loaded event such as a defeat may be the result of an outside intervention. In this case, we may attribute it to "bad luck"—or use the American vulgar slang "shit happens." Some bad things happen for no specific reason, but this doesn't mean that the event has no reason in the causal sense. A person might break his neck as a result of a highly improbable event, such as a heavy pelican experiencing a heart attack and falling from the sky onto this poor "victim of Lady Fortuna" who was "just trying to enjoy a vanilla ice cream on the beach," as the press might retrospectively describe the event. This event has a clear reason. The death of the poor victim was the result of a heavy pelican falling on him. Saying that "shit happens" doesn't mean that no cause is detected, just that negative events happen that are not personally targeted against the unlucky person who experiences the event. In some cases, the chain of events leading to an improbable event may be long and complex. Such a chain of events is illustrated in the hilarious comedy *Zig Zag Story* (1983),[2] directed by Patrick Schulmann. However, even in these complex chains, causality is evident. Still, nothing is personal in the sense that we cannot identify an intentional "agent," whether human or supernatural, that bears responsibility for the negative outcome.

So, shit happens, but can we use this vulgar phrase to describe what happened to the Brazilian soccer team? Most of us would not consider Brazil's defeat in such terms. It seems that the Brazilian team somehow collapsed from *within* as a result of German pressure. If this was the case, then we should try to understand how to approach such breakdowns. To start our discussion, we should return to *Anna Karenina*.

In many cases, we may hypothesize that a rare outcome, whether a positive or a negative one, is the result of a *unique* combination of constituting elements, similarly to the unique combination of letters forming a rare word. As mentioned in Chap. 1, in his opening to the novel *Anna Karenina*, Tolstoy (1877/2002) writes that "all happy families are alike" but that "each unhappy family is unhappy in its own way." Depending on the rarity of happy or unhappy families in each society, we may hypothesize that an observable thing (e.g. a happy family) that is conceived by us as rare can be explained by an underlying and unique *combination of components*. If being a happy family is the exception rather than the rule, then to be a happy family, we must have a rare combination of several features, such as financial stability, physical and mental health among all family members, and so on. *Ipso facto*, the more common combination is the one requiring fewer components.

The "Anna Karenina principle"[3] suggests that a deficiency in any one of a number of components constituting a rare event (e.g. being a happy family) dooms the event to failure. Consequently, a "successful endeavor" is one where every possible deficiency of a component has been avoided. This idea explains why some events, such as the existence of a happy family, are so rare, at least according to Tolstoy. To emerge and be sustained, these events *by necessity* require the combination of several components. However, if the defeat of the Brazilian football team was this kind of rare event, does it mean that the Brazilians lacked some components essential for winning or avoiding a shameful defeat? Or that the German team was exceptional in expressing a combination of features required to lead them to heroic victory? The answer seems to be positive but non-informative. Indeed, the Brazilians lacked a functioning defense system. However, this is something anyone observing the

---

[2] https://www.tcm.com/tcmdb/title/480690/zig-zag-story#overview.

[3] https://en.wikipedia.org/wiki/Anna_Karenina_principle#:~:text=The%20Anna%20Karenina%20principle%20states,possible%20deficiency%20has%20been%20avoided.

game could clearly see. To recall (see Chap. 1), in order to understand small social systems, we must go beyond grandmothers' intuition.

The Anna Karenina principle seems to be highly reasonable, specifically in the context of small social systems. To achieve a happy family, there are several criteria that must be met. Even a single violation of these criteria might lead to the destruction of the family's happiness. Think, for example, about family dramas where a father is missing, such as *The Glass Menagerie* by Tennessee Williams (1944/2011). Is it surprising that the missing father is one source of the family's misery? Or think about a soccer team where the top talent is missing from the match. Is it surprising that the team doesn't perform well? Or even imagine a jazz trio where one of the players is absent minded during a session and "diverges" from the groove. Is it surprising that in this context the trio doesn't play well?

We must admit that the Anna Karenina principle has great intuitive appeal. It actually says something quite simple: for a rare event to happen, a unique combination of "components" is a must in the most basic and logical sense of the combination. You must have component A AND component B AND component C in order for the event to happen. Here, the connective AND is used in the strict logical sense. In order for the whole expression (e.g. A AND B) to be true, *both* A AND B must be true.

Let's illustrate this point using a simple example. The sentence "I'm Yair Neuman and I'm writing this book" is true as long as the two arguments composing it:

(1) I'm Yair Neuman
(2) I'm writing this book

are both true. Even if one of the arguments is false, that's enough to dismiss the whole sentence as false. This is how logic works. If you have studied logic, then you will remember that AND is very strict about the way in which the truth-value is determined. The Anna Karenina principle suggests something that is in line with the spirit of logic. To happen, a rare event, such as being a happy family, must have *all* the required components. This principle is unforgiving as it makes the rare event far from resilient. Even a single violation of the AND logical gate is enough to destroy the unique and low-probability structure of "happy families," for instance.

We may easily understand this point by thinking about musical bands. The Beatles could not have been The Beatles without John Lennon AND Paul McCartney. Queen could not have been Queen without Freddie Mercury AND Brian May. The WHO could not have been the WHO without Roger

Daltrey AND Pete Townshend. Happy families seem to be no different from happy bands. However, returning to the soccer analogy, here comes the catch. It is more difficult to be a wining soccer team than a losing soccer team, but a defeat is different from just losing a game. A happy family can live a happy life with a missing father or if its economic situation is not the best. Suffering the misery of poverty is clearly a barrier to happiness but here we get into the trap of a circular argument in which being happy is defined by not suffering misery and if a family is suffering the misery of poverty, then it is deductively implied that it is not a happy family. Moreover, the issue is not whether it is *either* a happy family *or* an unhappy family. When Brazil played against Germany, it was a "happy" team, but something happened to it during the match. Therefore, what we are considering is not an issue of a simple binary classification of happy versus unhappy families, nor is it a combinatorial puzzle where a small system (such as The Beatles) could not have existed without a unique combination of talents. We are talking about something that is much more delicate than the binary classification of systems.

The combinatorial logic underlying the Anna Karenina principle is appealing but cannot be taken at face value. Adaptive systems (such as the immune system, soccer teams, or families) are not just simple combinations of components, and the behavior of adaptive systems doesn't simply follow the logic of the truth-functional operator AND. An adaptive system cannot allow itself to work along the unforgiving lines of the truth-functional operator AND. The existence of small social systems is so fragile and far from equilibrium that errors must be forgiven, deviations must be corrected, and failures must be overcome. Otherwise, both our genome and soccer teams would be easy prey for the annihilating tendency toward maximum entropy. Meeting the high bar of the Anna Karenina principle is impossible for an adaptive system. Therefore, and except in rare and extreme cases, the Anna Karenina principle seems to have a limited explanatory power.

As insightful argued by Rashevsky (1955), the complexity of natural observables cannot be explained simply by the limited variety of their components—and, let me add, their more or less unique combinations. It seems that the most important information about the behavior of a small social system exists in between the components and therefore that a catastrophic and rare event (such as the defeat of the Brazilian football team) should be attributed to the inner dynamic of the small system and the way it responds to a "disturbance" (in this case, the one imposed by the German team).

While some conditions must be met in order to gain a positive and relatively rare outcome, such as being a happy family, it seems that in small social systems, such as families and soccer teams, the source of success and failure may be deeply grounded in the inner dynamic of the team and the way it organizes itself to address internal and external challenges. Think, for example, about the idea of the weakest link in a chain. According to this common wisdom, the strength of a system is not determined by the sum of its components' strengths or by their average strength but by the weakest link in the chain. Recall Achilles, the famous Greek hero. When Achilles was a baby, his mother took him to the River Styx, which was supposed to make him invulnerable. According to this story, Achilles' mother was caring but clearly not careful, as she held Achilles by his heel to dip him into the water but then forgot to dip his heel afterwards. In fact, the story is interesting in terms of the mother's personality (a topic seldom discussed). In trying to save her son from his destiny, the mother makes the effort of dipping the baby into the shielding holy water but behaves in such a sloppy way that the baby is left with a vulnerable spot. She is a caring mother, but possibly also a mother suffering from attention deficit disorder. As a result of his mother's caring treatment, Achilles' body becomes invulnerable with the exception of one soft spot, the heel. According to one version of the story, which I still recall from my childhood, the "invulnerable" Achilles loses his life when an arrow shot by his opponent, Paris, hits him exactly in that soft spot. On average, Achilles, was highly invulnerable, but this average superiority did not take account of his Achilles' heel. As you may recall from previous discussions, the average is not always the best measure for understanding a given system.

How should we judge the strength of Achilles? According to the expression "a chain is only as strong as its weakest link," Achilles' strength was far below that suggested by his mythical reputation. A similar idea is proposed by *Liebig's law of the minimum*, which suggests that growth is not simply limited by the resources available but by the *scarcest resource* (limiting factor). Think about a baby whose growth is limited by a lack of vitamin B1. In 2003, a baby plant-based formula manufactured by the German company Humana and sold in Israel resulted in severe vitamin deficiencies in babies, some of whom were hospitalized with cardiac and neurological symptoms. "Three of them died, and at least twenty others were left with severe disabilities."[4] The lack of a *single* vitamin was a limiting factor in the babies' normal and healthy growth. Being aware of Liebig's law and its potentially devastating consequences might have led Humana to be more careful about the production and monitoring of its product.

In some cases, where growth or behavior is constrained by a limiting factor, a system can self-organize to compensate for the weakness of a certain link,

---

[4] https://en.wikipedia.org/wiki/Infant_formula.

but in other cases a limiting factor can have deadly consequences. Is a family the same as an infant's vulnerable developing brain? Can a successful soccer team be described similarly to a successful rock band? In certain respects, the answer is positive. However, as a result of learning the lesson of Leicester (see Chap. 2), our focus is on the surprising whole, which is different from the simple sum and combination of its components.

Following our entropy-based analysis of football (see Chap. 5), we can think about the Brazil–Germany match in terms of the weakest link in the chain (e.g. the defensive players) being broken under pressure, leading (through a positive feedback loop of escalating stress) to an increase in entropy, a crack-up of order, and deterioration to a tipping point where the team finally collapsed.

We can also think of a scenario where the weakest point is broken but there is no way of containing the break, such as if a hole opens up in a ship sailing in troubled water with no one to fix it. The positive feedback loop may lead to an accelerated chain reaction with a super-exponential growth function describing a failure that cannot be controlled. As the chain reaction unfolds, the entropy of the system increases, leading to the breakdown of order, communication, and control. There seems to be no way to stop this flood, and the small system (where information exists in the topology of the network) becomes an aggregate of components that have no functional connectivity or synergistic product as a working whole. The "team" stops behaving like a functional whole. To understand the importance of fixing a hole in a boat traveling in troubled waters, we may turn to the heroic story of the *Endurance*.

"By Endurance We Conquer"[5]

In a world almost completely mapped by human beings and their advanced technologies, we may find it hard to imagine a period when it was not just the case that some territories were unknown (*terra incognita*) but there were also considered to be territories where *hic sunt leones* (literally "here are lions")—unknown and dangerous territories existing beyond the reach of human civilization. Antarctica, the southernmost continent of the earth, was one of these territories. In 1914, an expedition led by a famous explorer, named Ernest Shackleton, traveled to Antarctica with the mission of crossing the continent from sea to sea.

---

[5] Sir Ernest Shackleton's family moto.

Shackleton's first-person memoir of the expedition reads like a breathtaking thriller (Shackleton, 1919/1999). It presents the struggle for survival experienced by Shackleton and his team after their ship—*Endurance*—was caught in an ice flue and damaged to a degree where the 28-man crew was ultimately forced to abandon it. The team initially survived on the ship and then moved onto an ice pack. Subsequently, using small boats, they sailed to Elephant Island and after 16 months set foot on solid land. However, the odyssey continued when Shackleton and five of his men risked themselves in an open-boat journey of three month to try to get help from a whaling station located on the South Atlantic island of South Georgia. Arriving at South Georgia, Shackleton and two of his men—Tom Crean and Frank Worsley—began hiking on foot to find the whaling station. After arriving at the station, it took them three additional months to rescue the rest of the team from Elephant Island and almost 21 months to once again set foot within human civilization.

The challenges the team had to cope with during this odyssey are unimaginable: hunger and starvation, freezing cold, and the danger of insanity were constant guests, and death was around every corner. Reading about the dangers, challenges, risks, and stress to which these men were exposed until they were rescued, it is almost impossible to believe that they survived against all the odds, having been pushed to the limit of human endurance.

Through the survivors' testimonies, we learn that Shackleton's personality was a major factor in the survival of the team. First, he had his own unique way of selecting the team members in a period when psychological testing didn't exist as an academic practice. His careful selection of the crew has been proved to have resulted in the selection of the fittest. He was also a highly charismatic figure and a leader his people fully trusted. Shackleton was highly sensitive to the dynamic of the team and responded to it in a way that, at least according to his own reported and biased judgment, was crucial for their survival. This personal account is supported by the testimonies of the other crew members as well.[6] Shackleton was in fact so sensitive to the psychological dynamic of his team that he refused to hoard food for the future because he was afraid that such a move might be interpreted as a sign of despair. His refusal to hoard food seemed irrational to some of his team members. However, his deep understanding of the dynamic of a group under stress led him to a counterintuitive choice that in retrospect seems to have been fully justified. Shackleton's heroic leadership was a crucial factor in the survival of the team, but it cannot be dissociated from the group's dynamic.

---

[6] See the two documentary films https://www.youtube.com/watch?v=HuKUX-7x2Ws and https://www.youtube.com/watch?v=sgh_77TtX5I.

Through Shackleton's own perspective and the testimonies of others, we may learn about an extraordinary group dynamic that led to the almost impossible survival of most of the team.

The Brazilian football team could have analyzed and studied the heroic struggle of the *Endurance* team as a lesson. By any objective measure, the hardship experienced by Shackleton and his men was incomparable to that experienced by the Brazilians playing against the German football team. However, the Brazilians cracked up while Shackleton and his men did not; on the contrary, they showed fighting spirit that was a model of courage, hope, and resilience. What can we learn from the first-person testimony of Sir Ernest Shackleton?

In the preface of his book, Shackleton admits that although he failed in his original mission to cross Antarctica, his book is worth reading as it records unflinching *determination*, supreme *loyalty*, and generous *self-sacrifice*. We may ask whether determination, loyalty, and self-sacrifice were the leading characteristics of the team's survival and answer this question through reading the book. Determination, or persistence, has been a recognized virtue since antiquity and one of the ideals characterizing the stoic. If they had not been determined, Shackleton's team would have fallen into despair, which is detrimental to survival. Loyalty and self-sacrifice are two closely related virtues expressed in the well-known motto of the Three Musketeers (Dumas, 1844/2007): "All for one and one for all." It seems that the famous Three Musketeers were a successful triad of warriors for the same reasons that Shackleton's men were a successful team of survivors: they were determined, never gave in to despair (despite experiencing it in large amounts) and exhibited self-sacrifice and loyalty. However, there is nothing new about these virtues. Any educated reader who has read enough classics is well aware of these virtues. It seems that "experts" in leadership and group dynamics who market these virtues make their living only through our own ignorance and oblivion. The classics and the insights we can gain through analysis of case studies such as the one of the *Endurance* are rich sources for understanding a team's resilience. What is missing from our understanding is a scientific approach through which the delicate and hidden aspects of a team's dynamic may be elucidated.

On the Poverty of Plans and the Importance of Absorbing a Punch or Two

In his book, Shackleton describes the careful plan of the expedition. As they brought with them sled dogs, he even carefully estimated the dogs' ability to "cover fifteen to twenty miles a day with loaded sledges" (Shackleton, 1919/1999, p. 3). For an Antarctic expedition, careful planning seems to be a must, just as it is for a soccer team. However, this planning must be accompanied by a deep understanding that every general plan may become irrelevant in given circumstances and may require real-time adjustments. As argued by Mike Tyson, a boxer mentioned in Chap. 2, "Everybody has a plan until they get punched in the mouth." When asked about the context of this well-known quotation,[7] Tyson said that he usually disregarded the plans and styles of his opponents. He explained that when a plan encounters a barrier, such as a surprising punch, the opponent may experience a paralyzing shock. This is the moment when the plan collapses and the question is, "What now?" I'm not sure whether the Brazilians playing against the German football team had a well-defined plan, but what we know for sure is that even if they had one, they cracked up when they were "punched" by the Germans twice in the opening quarter of the match.

Being able to respond in a flexible manner and in real time is the main virtue that I have attributed to small social systems. In the context of Shackleton's expedition, the loss of their ship—the *Endurance*—required a flexible and real-time response. Shackleton describes this event in sentimental terms. To a sailor, he says, a ship is more than a floating home (Shackleton, 1919/1999, p. 82), and he further describes the ambitions, hopes, and desires associated with his ship. Giving up your dreams may be a traumatic event. However, his and his team's flexibility are expressed in Shackleton's worldview that "a man must shape himself to a new mark directly the old one goes to ground" (p. 85). Indeed, the first expression of Shackleton's flexibility was in simply giving up his original plan to cross the continent and replacing it with the mission of saving the life of his people. This adjustment should not be underestimated. Shackleton was a very ambitious person who was seeking glory in his attempt to cross the continent. In fact, given his ambitious personality and his strong ego, we should particularly appreciate his flexibility in giving up his dream and changing the mission to one of survival.

---

[7] https://www.sun-sentinel.com/sports/fl-xpm-2012-11-09-sfl-mike-tyson-explains-one-of-his-most-famous-quotes-20121109-story.html.

This flexible adjustment of the expedition's plans didn't happen without a protest. One of the weakest links in the chain, according to some testimonies, was the carpenter Harry "Chippy" McNish, who had a bitter dispute with his captain after they lost the *Endurance*. Although opinions diverge between the testimonies, the fact that Shackleton was able to bypass Chippy's rebellion without executing him (a likely response in those days) or leaving him alone on the ice is a mark of flexibility and testament to Shackleton's ability to handle a weak link in the chain.

Flexibility and adjustment were not the only factors in the survival of Shackleton and his crew. We have previously discussed the importance of chemistry, and we cannot understand this amazing survival story without taking into account the chemistry between Shackleton and his second-in-command, Frank Wild. To the recipe of this survival story, we should also add the ingredient of synergy (again, discussed before). Although bad blood existed between Shackleton and his antagonistic carpenter, McNish, when he assembled the "elite force" who joined him in the open-boat journey to South Georgia after the loss of the *Endurance*, Shackleton chose McNish as one of the members. The crew seemed to be able to transcend particular and personal frictions. Adding a plus and a minus like Shackleton and McNish should mathematically lead to a neutral result, but in practice it produced a synergistic and positive result.

Given the above case study and our insights, we may now move on to the next phase and ask (1) how to identify the weakest link in the chain and (2) how to measure the way in which the system flexibly organizes around its weakness. First, let's measure the relative contributions of each player to each of the subgroups (e.g. triads) in which he can be assigned. Given the importance of triads in our understanding of small systems, we may specifically measure the relative contributions of the individual players to a triad of players, and the relative contributions of the players to the entropy of the ball passes. We start by examining a set of soccer matches and for each football match we measure the entropy of ball passes characterizing each triad of players. Then, we apply the idea of the Shapley value, use machine learning, and measure the relative contributions of each player across all triads and matches in which he participates. At this point, we may know something about the relative contributions of each player, and we may move on to the "test phase."

In the test phase, we look for matches in which the team has experienced a clear crisis. In the context of soccer, the simplest form of crisis is a goal conceded by the team. More complex crises are goals conceded by the team and goals that negatively increase the goal difference. If our team does not experience a crisis during a match, then it is a perfectly happy family that deserves jealousy but nothing more. However, if the team experiences a crisis during a match, we mark this time point and use it to segment the match into different time-intervals. Using this simple procedure, we may segment the match and measure whether the appearance of a crisis is expressed in the behavior of the players and their relative contributions, and we examine which player negatively responds to this crisis more than the others.

The simplest case is a single goal conceded by a team. The goal conceded by the team is a point of crisis, and we may compare the behavior of the team before and after the crisis takes place. The weak link in the chain is exposed by the change in (1) the relative contribution of a player and (2) the relative contributions of the team members to the synergy measure of the team as a whole or one of its subgroups. To measure the change in the player's contribution to the synergy of a subgroup (e.g. a triad), we may first apply the synergy measurement procedure as described in the previous chapter. Next, we measure the relative contributions of each player to the synergy, before and after the crisis.

A weak link is a player whose relative contribution significantly drops after a crisis has been experienced. A team suffers from suboptimal resilience if a change in the relative contribution of an individual player is expressed in a sub-additive behavior of the team as a whole. A resilient team is a team that is not influenced (or influenced to a minor degree) by a change in the behavior of its weakest link. Using this simple modeling procedure, we may identify the weakest link in the system and the way in which a team self-organizes after a crisis. When we read about the crises experienced by the *Endurance* team, we may be amazed by the marginal impact of the weakest links in the chain. We must admit that the weakest links in Shackleton's team were incredibly strong from the beginning, although relative to others there were people who were weaker than others. The team's dynamic was such that the more severe effect of the crisis on its weakest links didn't propagate to the other members to the extent of causing the synergy of this heroic team to "crack up."

Identifying the weakest link in a team and the subsets of players who are highly sensitive to the weakest link and to a crisis may help us to learn important lessons and prepare the team for the future. To the best of my knowledge, such a modeling procedure as described above does not currently exist and is not in use. While the proposed modeling procedure is illustrated in the context of soccer, it can be applied to any kind of small social system. Identifying the weakest links in a chain and the unique dynamic that allows them to be absorbed or neutralize may make an important contribution to the understanding of small social system and to our endurance.

Summary

- A crisis experienced by a small system may be attributed to the existence of a weak link.
- A weak link is a player whose relative contribution significantly drops after a crisis has been experienced.
- A team suffers from suboptimal resilience if a change in the relative contribution of an individual player is expressed in the behavior of the whole team.
- The resilience of the team may be expressed through its flexibility in addressing a weak link.
- Understanding a small social system may benefit from a methodology for identifying a weak link and understanding how to strengthen the link and the team's response to it.

Epilogue. Beating Death, Sometimes

"You can't beat death but / you can beat death in life, sometimes." (Charles Bukowski, "The Laughing Heart")

In Tractate Avot,[1] a Jewish sage by the name of Akavia ben Mahalalel suggests that we should keep in mind three things:

1. "From where you came,"
2. "To where you are going," and
3. "before Whom you will have to give an account and a reckoning."

Answering the first question ("From where did you come?"), he says, "From a putrid drop" (i.e. a drop of sperm). Answering the second question (i.e. "And to where are you going?"), he says, "To a place of dust, worms, and maggots." The Jewish sage reminds us that the miracle of life, as expressed in the existence of a whole human being, emerges from a drop of sperm and will finally end inmaximum entropy, as predicted by the second law of thermodynamics. Ashes to ashes, dust to dust. The lesson taught by the sage is that our arrogance and self-importance must be qualified. All human beings, whether kings or beggars, originated from the same source and are heading toward the same end. No one is a superman and in the world to come, we will all be responsible for giving "an account" of our deeds.

---

[1] https://en.wikisource.org/wiki/Translation:Mishnah/Seder_Nezikin/Tractate_Avot/Chapter_3/1.

Y. Neuman, *How Small Social Systems Work*, The Frontiers Collection, https://doi.org/10.1007/978-3-030-82238-5

Interestingly, the sage attributes our misbehavior to our misunderstanding of who we are: cognate beings who originated from matter and are heading back to a lower-level organization of matter. The need for modesty implied by this lesson is important but so is the understanding that being alive means being able to maintain a functional and organizational whole. This is a challenge not only because an annihilating force always lies at our doorstep, striving to return us to our origin, but also because "learning order" seems to be a challenge in an irreversible world where going back in time is impossible. Learning is impossible for a being living in a continuous present. To learn, we must go back in time.

At the opening of this book, I proposed that our existence is cognitive no less than it is physical because without a mind capable of registering the past and forming abstract representations, existence is impossible. However, I have also argued that our existence is social no less than it is physical and cognitive and that the formation of flexible and adaptive representations and the ability to respond flexibly and creatively in real time is what underlies the existence of small social systems. My main argument was that a small number of "cognitive particles" may hold different perspectives that, when compared and integrated, may have a huge benefit. It is our in-built differences in perspective and our ability to somehow resolve these non-converging perspectives which create the magical sauce of small social systems from soccer teams to families. As suggested by Charles Bukowski in his poem "The Laughing Heart," (Bukowski, 1996) we can't beat death but we can beat death in life, sometimes. When a soccer team is playing beautifully, forming complex, dynamic configurations of order, it is an example of the way we beat death in life, sometimes. When Scott Hamilton's quartet[2] beautifully improvises, it is another instance of the way in which we may sometimes beat death in life. When a functioning family is caring for its members, from infants to grandparents, life has the upper hand. These instances emphasize the importance of small social systems that have been overlooked, misunderstood, and compartmentalized into specific niches of academic disciplines. In between the socialist family of nations and Ayn Rand's form of extreme individualism, there seems to be a level of analysis that deserves careful consideration, a level where a constructive resolution is formed between the understanding that we are all unique and the understanding that we are never alone.

---

[2] https://www.youtube.com/watch?v=SMFfXqt2VZo&list=WL&index=21.

References

Alemi, M. (2020). *The amazing journey of reason: From DNA to artificial intelligence.* Springer.

Aleta, A., & Moreno, Y. (2019). Multilayer networks in a nutshell. *Annual Review of Condensed Matter Physics, 10*(1), 45–62.

Alexander, C. (2002). *The nature of order: The process of creating life.* Taylor & Francis.

Ashby, W. R. (1991). Requisite variety and its implications for the control of complex systems. In G. J. Klir (Ed.), *Facets of systems science* (pp. 405–417). Springer.

Atlan, H., & Cohen, I. R. (1998). Immune information, self-organization and meaning. *International Immunology, 10*(6), 711–717.

Bakhtin, M. M. (1990). In M. Holquist & V. Liapunov (Eds.), *Art and answerability: Early philosophical essays* (Vol. 9). University of Texas Press.

Bateson, G. (1972/2000). *Steps to an ecology of mind: Collected essays in anthropology, psychiatry, evolution, and epistemology.* University of Chicago Press.

BBC. (2014). Brazil "cracked up" during Tuesday's stunning 7–1 World Cup semi-final defeat by Germany, according to victorious coach Joachim Low. *BBC Sport,* July 9. https://www.bbc.co.uk/sport/football/28222946

Ben-Jacob, E., Becker, I., Shapira, Y., & Levine, H. (2004). Bacterial linguistic communication and social intelligence. *Trends in Microbiology, 12*(8), 366–372.

Bennett, C. H., & Landauer, R. (1985). The fundamental physical limits of computation. *Scientific American, 253*(1), 48–57.

Billig, M. (1987). *Arguing and thinking: A rhetorical approach to social psychology.* Cambridge University Press.

Bukowski, C. (1996). *The laughing heart.* Black Sparrow Press.

Y. Neuman, *How Small Social Systems Work*, The Frontiers Collection, https://doi.org/10.1007/978-3-030-82238-5

Borges, J. L. (2018). *The garden of forking paths*. Penguin.

Bransen, L., & Van Haaren, J. (2020). *Player chemistry: Striving for a perfectly balanced soccer team*. arXiv preprint arXiv:2003.01712.

Brontë, E. (1847/2020). *Wuthering Heights*. Oxford University Press.

Cohen, I. R. (2000). *Tending Adam's garden: Evolving the cognitive immune self*. Academic.

Cohen, I. R. (2016). Updating Darwin: Information and entropy drive the evolution of life. *F1000Research, 5*, 2808.

Collins, J., Howe, K., & Nachman, B. (2018). Anomaly detection for resonant new physics with machine learning. *Physical Review Letters, 121*(24), 241803.

Conant, R. C., & Ross Ashby, W. (1970). Every good regulator of a system must be a model of that system. *International Journal of Systems Science, 1*(2), 89–97.

Conway, M. A., Loveday, C., & Cole, S. N. (2016). The remembering–imagining system. *Memory Studies, 9*(3), 256–265.

Danesi, M. (2004). *Messages, signs, and meanings: A basic textbook in semiotics and communication* (Vol. 1). Canadian Scholars' Press.

Darwin, C. (1859/2010). *The origin of species: A variorum text*. University of Pennsylvania Press.

Davies, P. C., & Walker, S. I. (2016). The hidden simplicity of biology. *Reports on Progress in Physics, 79*(10), 102601.

Dumas, A. (1844/2007). *The three musketeers*. Dover.

Farzadfard, F., Gharaei, N., Higashikuni, Y., Jung, G., Cao, J., & Lu, T. K. (2019). Single-nucleotide-resolution computing and memory in living cells. *Molecular Cell, 75*(4), 769–780.

Fortuny, J., & Corominas-Murtra, B. (2013). On the origin of ambiguity in efficient communication. *Journal of Logic, Language and Information, 22*(3), 249–267.

Fredkin, E., & Toffoli, T. (1982). Conservative logic. *International Journal of Theoretical Physics, 21*(3–4), 219–253.

Friedgut, E., & Bourgain, J. (1999). Sharp thresholds of graph properties, and the k-sat problem. *Journal of the American Mathematical Society, 12*(4), 1017–1054.

Fu, X., Kato, S., Long, J., Mattingly, H. H., He, C., Vural, D. C., Zucker, S. W., & Emonet, T. (2018). Spatial self-organization resolves conflicts between individuality and collective migration. *Nature Communications, 9*(1), 1–12.

Galton, F. (1889). *Natural inheritance*. Macmillan.

Gigerenzer, G. (1997). Bounded rationality: Models of fast and frugal inference. *Swiss Journal of Economics and Statistics, 133*(2/2), 201–218.

Gladwell, M. (2013). *David and Goliath: Underdogs, misfits, and the art of battling giants*. Little, Brown.

Hardison, O. B. (1960). The dramatic triad in "Hamlet." *Studies in Philology, 57*(2), 144–164.

Harries-Jones, P. (1995). *A recursive vision: Ecological understanding and Gregory Bateson*. University of Toronto Press.

Harries-Jones, P. (2016). *Upside-down gods: Gregory Bateson's world of difference*. Fordham University Press.

Hartmann, A. K., & Weigt, M. (2005). *Phase transitions in combinatorial optimization problems.* Wiley.

Hemingway, E. (1970/2013). *Islands in the stream.* Random House.

Hemmo, M., & Shenker, O. R. (2012). *The road to Maxwell's demon: Conceptual foundations of statistical mechanics.* Cambridge University Press.

Hiraide, T. (2014). *The guest cat.* Picador.

Holmes, L. (2016). Resolution of missing incidents. In K. Shalev Greene & L. Alys (Eds.), *Missing persons: A handbook of research* (pp. 234–242). Routledge.

Jaglom, H. (1992). The independent filmmaker. In J. E. Squire (Ed.), *The movie business book* (2nd ed., pp. 49–59). Simon & Schuster.

Jaynes, E. T. (1957). Information theory and statistical mechanics. *Physical Review, 106*(4), 620–630.

Jaynes E. T. (1996/2003). *Probability theory: The logic of science.* Cambridge University Press.

Keim, N. C., Paulsen, J. D., Zeravcic, Z., Sastry, S., & Nagel, S. R. (2019). Memory formation in matter. *Reviews of Modern Physics, 91*(3), 035002.

Komikova, M. (2020). *The biggest bluff.* Penguin.

Kurt, W. (2015). Using Bayes' factor to build a Voight-Kampff test! *Count Bayesie,* May 5. https://www.countbayesie.com/blog/2015/2/27/building-a-bayesian-voight-kampff-test

Lan, G., & Tu, Y. (2016). Information processing in bacteria: Memory, computation, and statistical physics—A key issues review. *Reports on Progress in Physics, 79*(5), 052601.

Leff, H. S., & Rex, A. F. (Eds.). (2014). *Maxwell's demon: Entropy, information, computing.* Princeton University Press.

Long, Z., Quaife, B., Salman, H., & Oltvai, Z. N. (2017). Cell–cell communication enhances bacterial chemotaxis toward external attractants. *Scientific Reports, 7*(1), 1–12.

Lutz, E., & Ciliberto, S. (2015). From Maxwell's demon to Landauer's eraser. *Physics Today, 68*(9), 30–35.

Marcelino, R., Sampaio, J., Amichay, G., Gonçalves, B., Couzin, I. D., & Nagy, M. (2020). Collective movement analysis reveals coordination tactics of team players in football matches. *Chaos, Solitons & Fractals, 138,* 109831.

Martin, O. C., Monasson, R., & Zecchina, R. (2001). Statistical mechanics methods and phase transitions in optimization problems. *Theoretical Computer Science, 265*(1–2), 3–67.

Middleton, D. E., & Edwards, D. E. (Eds.). (1990). *Collective remembering.* Sage.

Miller, G. A. (1956). The magical number seven, plus or minus two: Some limits on our capacity for processing information. *Psychological Review, 63*(2), 81–97.

Mínguez-Toral, M., Cuevas-Zuviría, B., Garrido-Arandia, M., & Pacios, L. F. (2020). A computational structural study on the DNA-protecting role of the tardigrade-unique Dsup protein. *Scientific Reports, 10*(1), 1–18.

Mischel, W. (2004). Toward an integrative science of the person. *Annual Review of Psychology, 55*(1), 1–22.

Molenaar, P. C. (2013). On the necessity to use person-specific data analysis approaches in psychology. *European Journal of Developmental Psychology, 10*(1), 29–39.

Molenaar, P. C., & Campbell, C. G. (2009). The new person-specific paradigm in psychology. *Current Directions in Psychological Science, 18*(2), 112–117.

Neuman, Y. (2004). Meaning-making in the immune system. *Perspectives in Biology and Medicine, 47*(3), 317–327.

Neuman, Y. (2017). *Mathematical structures of natural intelligence*. Springer.

Neuman, Y. (2020). *Conceptual mathematics and literature: Toward a deep reading of texts and minds*. Brill.

Neuman, Y., Cohen, Y., & Neuman, Y. (2019). How to (better) find a perpetrator in a haystack. *Journal of Big Data, 6*(1), 9.

Neuman, Y., Israeli, N., Vilenchik, D., & Cohen, Y. (2018). The adaptive behavior of a soccer team: An entropy-based analysis. *Entropy, 20*(10), 758.

Neuman, Y., Lev-Ran, Y., & Erez-Shalom, E. (2020). Screening for potential school shooters through the weight of evidence. *Heliyon, 6*(10), e05066.

Neuman, Y., & Vilenchik, D. (2019). Modeling small systems through the relative entropy lattice. *IEEE Access, 7*, 43591–43597.

Neuman, Y., Vilenchik, D., & Cohen, Y. (2020). From physical to social interactions: The relative entropy model. *Scientific Reports, 10*(1), 1–8.

Noble, D. (2008). *The music of life: Biology beyond genes*. Oxford University Press.

Noble, D. (2012). A theory of biological relativity: No privileged level of causation. *Interface Focus, 2*(1), 55–64.

Parrondo, J. M., Horowitz, J. M., & Sagawa, T. (2015). Thermodynamics of information. *Nature Physics, 11*(2), 131–139.

Peirce, C. S. (1897). The logic of relatives. *The Monist, 7*(2), 161–217.

Pennebaker, J. W., Boyd, R. L., Jordan, K., & Blackburn, K. (2015). *The development and psychometric properties of LIWC2015*. University of Texas at Austin.

Peters, O., & Gell-Mann, M. (2016). Evaluating gambles using dynamics. *Chaos: An Interdisciplinary Journal of Nonlinear Science, 26*(2), 023103.

Poirier, B. (2014). *A conceptual guide to thermodynamics*. Wiley.

Randall, D. J., Randall, D., Burggren, W., French, K., & Eckert, R. (2002). *Eckert animal physiology*. Macmillan.

Rashevsky, N. (1955). Life, information theory, and topology. *Bulletin of Mathematical Biophysics, 17*(3), 229–235.

Rathgeber, M., Bürkner, P. C., Schiller, E. M., & Holling, H. (2019). The efficacy of emotionally focused couples therapy and behavioral couples therapy: A meta-analysis. *Journal of Marital and Family Therapy, 45*(3), 447–463.

Reichardt, J., Alamino, R., & Saad, D. (2011). The interplay between microscopic and mesoscopic structures in complex networks. *PLoS ONE, 6*(8), e21282.

Risau-Gusman, S., Martinez, A. S., & Kinouchi, O. (2003). Escaping from cycles through a glass transition. *Physical Review E, 68*(1), 016104.

Sebeok, T. A., & Danesi, M. (2012). *The forms of meaning: Modeling systems theory and semiotic analysis* (Vol. 1). Walter de Gruyter.

Seneca. (2004). *On the shortness of life*. Penguin.

Sevick, E. M., Prabhakar, R., Williams, S. R., & Searles, D. J. (2008). Fluctuation theorems. *Annual Review of Physical Chemistry, 59*, 603–633.

Shackleton, E. (1919/1999). *South: The* Endurance *expedition*. New York, NY: Penguin.

Shakespeare, W. (1987). In G. R. Hibbard (Ed.), *Hamlet*. Oxford University Press.

Shakespeare, W. (2004). *Romeo and Juliet*. Simon & Schuster.

Shannon, C. E. (1956). The bandwagon. *IRE Transactions on Information Theory, 2*(1), 3.

Shemesh, Y., Sztainberg, Y., Forkosh, O., Shlapobersky, T., Chen, A., & Schneidman, E. (2013). High-order social interactions in groups of mice. *Elife, 2*, e00759.

Shi, L., Beaty, R. E., Chen, Q., Sun, J., Wei, D., Yang, W., & Qiu, J. (2020). Brain entropy is associated with divergent thinking. *Cerebral Cortex, 30*(2), 708–717.

Simon, H. A. (1957). *Models of man*. Wiley.

Steinbeck. J. (1937/1993). *Of mice and men*. Penguin.

Taleb, N. N. (2012). *Antifragile: Things that gain from disorder* (Vol. 3). Random House.

Taleb, N. N. (2020). *Skin in the game: Hidden asymmetries in daily life*. Random House.

Tamir, B., & Neuman, Y. (2016). The physics of categorization. *Complexity, 21*(S1), 269–274.

Tavella, F., Giaretta, A., Dooley-Cullinane, T. M., Conti, M., Coffey, L., & Balasubramaniam, S. (2019). DNA molecular storage system: Transferring digitally encoded information through bacterial nanonetworks. *IEEE Transactions on Emerging Topics in Computing*, 1-1.

Tolkien, J. R. R. (1937/2012). *The hobbit*. Houghton Mifflin Harcourt.

Tolstoy, L. (1877/2002). *Anna Karenina* (R. Pevear & L. Volokhonsky, Trans.). Penguin.

Tolstoy, L. (1897/1995). *What is art?* Penguin.

van Es, T. (2020). Living models or life modelled? On the use of models in the free energy principle. *Adaptive Behavior*, 1059712320918678.

Volosinov, V. N. (1986). *Marxism and the philosophy of language*. Harvard University Press.

Wainer, H. (2016). *Truth or truthiness: Distinguishing fact from fiction by learning to think like a data scientist*. Cambridge University Press.

Weiss, G. (1975). Time-reversibility of linear stochastic processes. *Journal of Applied Probability, 12*(4), 831–836.

West, G. (2017). *Scale: The universal laws of growth, innovation, sustainability, and the pace of life in organisms, cities, economies, and companies*. Penguin.

Wilk, G., & Włodarczyk, Z. (2008). Example of a possible interpretation of Tsallis entropy. *Physica A: Statistical Mechanics and Its Applications, 387*(19–20), 4809–4813.

Williams, T. (1944/2011). *The glass menagerie*. New Directions.

Young, T. W. (2018). *The Sherlock effect: How forensic doctors and investigators disastrously reason like the great detective*. CRC Press.

Zipf, G. K. (2016). *Human behavior and the principle of least effort: An introduction to human ecology*. Ravenio Books.

Zu, P., Boege, K., del-Val, E., Schuman, M. C., Stevenson, P. C., Zaldivar-Riverón, A., & Saavedra, S. (2020). Information arms race explains plant–herbivore chemical communication in ecological communities. *Science, 368*(6497), 1377–1381.

Zuckerman, G. (2019). *The man who solved the market: How Jim Simons launched the quant revolution*. Penguin.

Author Index

Subject Index

Printed in the United States
by Baker & Taylor Publisher Services